2015 Estimation of Agricultural Productivity and Annual Update of Competitiveness Analysis for Greater China

NUS National University of Singapore

LKY Lee Kuan Yew School of Public Policy
ASIA COMPETITIVENESS INSTITUTE

In 2016 ACI was ranked 13th globally, 2nd in Asia and 1st in Singapore amongst 90 think-tanks world-wide under the "Best University Affiliated Think-Tank" category by Think-Tanks & Civil Societies Program at The University of Pennsylvania, USA

About ACI

The Asia Competitiveness Institute (ACI) was established in August 2006 as a research centre at the Lee Kuan Yew School of Public Policy (LKYSPP), National University of Singapore (NUS). It aims to build the intellectual leadership and network for understanding and developing competitiveness in the Asia region. ACI seeks to contribute to the enhancement of inclusive growth, living standards and institutional governance through competitiveness research on sub-national economies in Asia. It identifies mitigating issues and challenges for potential public policy interventions through close collaboration with regional governments, business corporations, policy think-tanks and academics. ACI's three key research pillars include (i) sub-national economies level competitiveness analysis and city-level liveability analysis, (ii) firm-level competitiveness analysis in 16 Asia economies, and (iii) Singapore's long-term growth strategies and public policy analysis.

ACI's value propositions may be encapsulated in its acronym:

Analytical inputs to initiate policies for policy-makers and business leaders in Asia

Capacity building to enable others through improvement in productivity and efficiency

Intellectual leadership to create pragmatic models of competitiveness and inclusive growth

The institute's core research competencies can also be encapsulated in this acronym describing our evidence-based assessments conducted on public policies for ASEAN in the context of the rise of China and India.

Vision and Mission

- ACI's over-arching vision is to build up its research credibility with policy impact, contributing as a professional, world-class think-tank.
- ACI's mission is to establish our niche as a leading policy think-tank by identifying competitiveness trends, opportunities and challenges, as well as promoting competition and synergising complementarities amongst Asian economies and business corporations.
- ACI endeavours to articulate sound recommendations, entice discourse, and shape agenda in the arena of public policy amongst Asian governments.
- ACI undertakes evidence-based analysis of public policy issues and decisions, in order to provide assessment of their effectiveness as well as economic and societal impact.

Research Initiatives and Collaborations

I. Identify trends of competitiveness and policy analysis on trade and investment of ASEAN, within the regional context of competition and complementarities with China and India.

II. Identify competitive strengths and conduct policy analysis on Singapore within the context of regional economies with international benchmarking.

III. We are regularly releasing three indices on liveability ranking including 64 Global Cities, 100 Greater China Cities and 17 Shandong Cities.

IV. We have established an Ease of Doing Business Index on Attractiveness to Investors, Business Friendliness and Competitive Policies for 21 states and federal territories of India and 33 provinces of Indonesia.

V. We have signed Memoranda of Understanding with The World Bank (2016 and 2015), SPRING Singapore (2014) and European Central Bank (2014).

VI. We have signed Memoranda of Understanding with various institutions in Greater China economies, including Shanghai Academy of Social Sciences (2016), Shandong Academy of Social Sciences (2015), Chongqing Municipal People's Government (2015), China Institute for Reform and Development, Haikou (2015), Counsellors' Office of the People's Government of Guangdong Province (LOI, 2014) and Chung-Hua Institution for Economic Research (2015), Taiwan.

VII. We have signed Memoranda of Understanding with six Chief Ministers' Offices in Andhra Pradesh (2015), Bihar (2015), Chhattisgarh (2015), Madhya Pradesh (2015), Odisha (2015) and Punjab (2015).

VIII. We have signed Memoranda of Understanding with Indonesia Investment Coordinating Board (2014), Indonesian Agency for Agricultural Research and Development at Ministry of Agriculture (2014), Indonesian President's Delivery Unit for Developing Monitoring and Oversight (2013) and Employer's Association of Indonesia (2013).

IX. We have signed Memoranda of Understanding with ASEAN think-tanks, including Centre for Strategic and International Studies, Indonesia (2015), Institute of Strategic & International Studies, Malaysia (2015), Philippine Institute for Development Studies (2015), Thailand Development Research Institute (2015) and Central Institute for Economic Management, Vietnam (2015).

About ACI's Research Pillars

ACI has consciously engaged in economic research that has significant relevance to Singapore and the Asian region. Over the years, ACI has focused on our expertise in quantitative competitiveness analysis and simulation, spinning off volumes of research output in the applications of our methodology and regional insight. At this stage, ACI has identified three core research pillars that will guide and define its research efforts moving forward. The three research pillars are as follows:

Pillar I. Sub-national Economies Competitiveness Analysis

ACI engages in systematic and methodical competitiveness analyses of the *sub-national economies* by using an evidence-based, empirical approach involving a comprehensive list of relevant indicators, which are categorised under multiple layers called the 'environments'. A number of our projects, including competitiveness analyses of Greater China, ASEAN-10, India, and Indonesia, as well as other thematic research projects, were conducted by adopting this common methodology, with potential variations in the specific environments and indicators. ACI's competitiveness analysis of Asian economies goes beyond the usual ranking to offer constructive policy recommendations on how individual member states can improve their rankings vis-à-vis their sub-national or regional peers through the application of the 'what-if' simulation, which provides the projected improvements of each state's ranking whereby the bottom one-fifth of its indicators are enhanced.

Pillar II. Micro-based Firm Level Competitiveness Analysis

Micro-based firm level competitiveness analysis is ACI's second research pillar that has been developed in view of the intrinsic importance of firm-level competitiveness in terms of productivity, efficiency, and governance. In the context of globalisation, mobility of economic activities, and blurring of borders, an understanding of the determinants and dynamics of firm-level competitiveness is paramount, in order for policy makers to make adjustments and prepare their industries for an increasingly competitive economic landscape. To this end, ACI has partnered with the European Central Bank to carry out research in this area, possessing a solid foundation in firm-level productivity research in the form of European Competitiveness Network database and methodology. ACI envisaged the expansion of CompNet into Asia, thereby pioneering the Asia CompNet research network encompassing 16 economies in Asia.

Pillar III. Singapore's Long-term Economic Growth Strategies and Public Policies Analysis

The third and imminent research pillar focuses on Singapore's long-term economic growth strategies in the context of changing circumstances, future trends, and emerging opportunities for Singapore in the decades ahead. In particular, ACI will take on the task of critically examining Singapore's public policy strengths and areas of improvement by assessing policy successes of the past, identifying new issues to address and the current policy gaps, through a systematic and evidence-based research inquiry using quantitative methodology and empirical data, leveraging on our network of policy experts.

Founding Patron and International Advisory Panel

Founding Patron: **Mr George Yeo**
Visiting Scholar, Lee Kuan Yew School of Public Policy, National University of Singapore & Former Minister of Foreign Affairs, Singapore

International Advisory Panel

Co-Chairs:

Professor Michael Porter (2006–2010)
Bishop William Lawrence University Professor

Ms Marjorie Yang (2006–2012)
Chairman, Esquel Group

Members:

Professor Kishore Mahbubani
Dean, Lee Kuan Yew School of Public Policy, National University of Singapore

Dr Kuntoro Mangkusubroto
Head, President's Delivery Unit for Developing Monitoring and Oversight (UKP4), Indonesia

Mr Narayana Murthy
Chairman, Infosys Technologies Limited

Mr Philip Yeo
Chairman, SPRING Singapore

Mrs Ow Foong Pheng
Permanent Secretary, Ministry of Trade and Industry

Ms Yong Ying-I
Permanent Secretary, Public Service Division

Professor Chan Kam Leung Alan
Dean, College of Humanities, Arts and Social Sciences,
Nanyang Technological University

Asia Competitiveness Institute

Co-Directors:

Associate Professor Tan Khee Giap (2011 till now)
Professor Tan Kong Yam (2011 till now)

Asia Competitiveness Institute - World Scientific Series

ISSN: 2424-9912

Series Editor: Tan Khee Giap (*Lee Kuan Yew School of Public Policy, NUS, Singapore)*

2014

2014 Annual Competitiveness Ranking and Simulation Study for ASEAN-10 and Development Strategies to Enhance Asia Economic Connectivity
by Tan Khee Giap, Sangiita Yoong Wei Cher, Sasidaran Gopalan & Nguyen Le Phuong Anh (Lee Kuan Yew School of Public Policy, NUS, Singapore)

2014 Annual Provincial and Regional Competitiveness Analysis for Greater China Economies: Development Strategies under the New Normal
by Tan Khee Giap, Yuan Randong & Sangiita Yoong Wei Cher (Lee Kuan Yew School of Public Policy, NUS, Singapore)

2014 Regional Competitiveness Analysis and a Master Plan on Regional Development Strategies for India: Annual Competitiveness Update and Evidence on Economic Development Model for Selected States of India
by Tan Khee Giap, Linda Low, Kartik Rao & Tan Kong Yam (Lee Kuan Yew School of Public Policy, NUS, Singapore)

2014 Provincial and Inaugural Regional Competitiveness Analysis: Safeguarding Indonesia's Growth Momentum
by Tan Khee Giap, Nurina Merdikawati, Mulya Amri & Tan Kong Yam (Lee Kuan Yew School of Public Policy, NUS, Singapore)

2014 Annual Competitiveness Analysis and Development Strategies for Indonesian Provinces
by Tan Khee Giap, Nurina Merdikawati, Mulya Amri & Blake Harley Berger (Lee Kuan Yew School of Public Policy, NUS, Singapore)

2015

2015 Annual Competitiveness Analysis, Agricultural Productivity and Development Vision for ASEAN-10
by Tan Khee Giap, Nguyen Le Phuong Anh, Sasidaran Gopalan & Luu Nguyen Trieu Duong (Lee Kuan Yew School of Public Policy, NUS, Singapore)

2015 Estimation of Agricultural Productivity and Annual Update of Competitiveness Analysis for Greater China: Optimising Agricultural Productivity and Promoting Innovation Driven Growth
by Tan Khee Giap, Yuan Randong & Xie Teleixi (Lee Kuan Yew School of Public Policy, NUS, Singapore)

2015 Agricultural Total Factor Productivity and Competitiveness Analysis for States and Federal Territories and Five Regions of India: Annual Competitiveness Update and Evidence on the Agricultural Development Models for Selected Indian States
by Tan Khee Giap, Sasidaran Gopalan & Anuja Tandon (Lee Kuan Yew School of Public Policy, NUS, Singapore)

2015 Agricultural Productivity, Decentralisation, and Competitiveness Analysis for Provinces and Regions of Indonesia
by Tan Khee Giap, Nurina Merdikawati, Mulya Amri & Tan Kong Yam (Lee Kuan Yew School of Public Policy, NUS, Singapore)

2015 Annual Competitiveness Analysis and Development Strategies for Indonesian Provinces
by Tan Khee Giap, Mulya Amri, Nurina Merdikawati & Nursyahida Binte Ahmad (Lee Kuan Yew School of Public Policy, NUS, Singapore)

The complete list of the volumes in the series published since 2013 can be found at
http://www.worldscientific.com/series/aciwss

Asia Competitiveness Institute - World Scientific Series

2015 Estimation of Agricultural Productivity and Annual Update of Competitiveness Analysis for Greater China

Optimising Agricultural Productivity and Promoting Innovation Driven Growth

Tan Khee Giap
Co-Director, Asia Competitiveness Institute
Lee Kuan Yew School of Public Policy, National University of Singapore

Yuan Randong
Research Associate, Asia Competitiveness Institute
Lee Kuan Yew School of Public Policy, National University of Singapore

Xie Teleixi
Research Assistant, Asia Competitiveness Institute
Lee Kuan Yew School of Public Policy, National University of Singapore

Reviewers

Zhang Shucun
Dean
Shandong Academy of Social Sciences, People's Republic of China

Chi Fulin
President
China Institute for Reform and Development (CIRD) & Member of the Chinese People's Political Consultative Conference, People's Republic of China

Chang Xiuze
Professor
Academy of Macroeconomic Research, National Development and Reform Commission, People's Republic of China

He Dongni
Vice President
China Institute for Reform and Development (CIRD), People's Republic of China

Ni Pengfei
Director
Centre for City and Competitiveness Chinese Academy of Social Sciences (CASS), People's Republic of China

Zhang Qingjin
Director
Institute of Rural Development Shandong Academy of Social Sciences, People's Republic of China

Lee Kuan Yew School of Public Policy
Asia Competitiveness Institute
NUS National University of Singapore
World Scientific

Published by

World Scientific Publishing Co. Pte. Ltd.
5 Toh Tuck Link, Singapore 596224
USA office: 27 Warren Street, Suite 401-402, Hackensack, NJ 07601
UK office: 57 Shelton Street, Covent Garden, London WC2H 9HE

British Library Cataloguing-in-Publication Data
A catalogue record for this book is available from the British Library.

Asia Competitiveness Institute - World Scientific Series
2015 ESTIMATION OF AGRICULTURAL PRODUCTIVITY AND ANNUAL UPDATE OF COMPETITIVENESS ANALYSIS FOR GREATER CHINA
Optimising Agricultural Productivity and Promoting Innovation Driven Growth

ISBN 978-981-3148-32-1

Desk Editor: Jiang Yulin

Typeset by Stallion Press
Email: enquiries@stallionpress.com

Printed in Singapore

Contents

Foreword

In an increasingly globalised world, economies need to stay open and competitive in order to stay relevant. And the state has an important role in setting policies that can nurture a country's competitiveness and improve the welfare of its people. After more than three decades of reform, China has experienced a momentous socio-economic transformation and emerged as the second largest economy and biggest trading nation in the world. China's growth and stability matter not only domestically, but also have a great impact on its neighbours in Asia as well as several other regions. Therefore, how the Chinese government devises and implements future development strategies has been closely watched. Will its future set of options remain largely the same? How much further will China go to bring in the foreign technologies, institutions and ideas that its people still need? To what degree will its government intervene and regulate an economy that could evolve towards a freer and more market-based one?

For answers to these questions, the findings and perspectives here may offer some help. The volume contains the annual tracking and monitoring by the Asia Competitiveness Institute (ACI) at the Lee Kuan Yew School of Public Policy (LKYSPP), National University of Singapore (NUS) of the evolving competitiveness in Greater China economies and regions since 2012. It is the third instalment of a series that provides an annual update of the competitiveness analysis of Greater China at both the provincial and regional levels covering the five regions constituting Greater China. Addressing regional imbalances in economic growth, a key aspect of the competitiveness landscape of Greater China, will be a deciding factor that will shape the country's future development. The research results on provincial and regional competitiveness as presented in this book will shed light on how to tackle regional disparities in development and their strategic implications.

Political philosophers in the West tend to invoke the dichotomy in systems of government when they theorize on the nature of nation states. Two diametrically opposing views of how a government functions are identified, the organic and the mechanistic, reminding us of a more traditional oriental-occidental divide. Such a division of ideas has a long tradition in the Judeo-Christian and Greco-Roman world which often considers anything in between, or beyond this divide, as uninformative or meaningless. Outliers found not to fit that Procrustean bed would need to be regularised and made to conform to their theories. Indeed, such a dichotomous approach has had an unduly heavy influence on the framework of thinking in the Western world. This is unfortunate as it casts a long shadow that darkens the present-day thinking of many in the West, the tendency to put things into boxes or categories. Functionally, it really does not matter whether a cat is black or white or any other colour as long as it catches mice, as the Chinese leader Deng Xiaoping once put it.

Dichotomous thinking can lead to over-simplifications. George W. Bush's view, "either you are with us, or you are with the terrorists", for example, although understandable given its context, led to other kinds of danger. Another example is what the United States and the United Kingdom did in Iraq after the invasion. Practising sectarianism and dividing Baghdad into exclusive zones of Sunnis and Shiites, with the administration backed by foreign military forces, completely changed the way the city was organised, paving the way for total disaster. Recently, China's "One Belt, One Road" strategy has been compared to the America's European Recovery Program known as the Marshall Plan. This is quite misleading. Underlying the Marshall Plan was the logic of opposing ideologies. China's proposal for international economic cooperation, however, is not aimed at making political friends and foes. Here it would be more realistic to understand the organisation of human societies as a continuum of views and methods and not through the division along ideological lines or by any single value system.

It is in this spirit of seeing the world as a continuum that ACI has adopted its comprehensive framework of competitiveness analysis. This consists of an eclectic yet relevant set of indicators that is used to gauge the dynamism of competitiveness factors at sub-national levels from an empirical, data-driven perspective. With the annual tracking and updating of such competitiveness analysis, the ACI presents a history of a modernising and developing China through sets of economic data. Now that China has entered into an era of the New Normal, where phenomenal growth in GDP is no longer the primary objective of various levels of the government, it is expected that the public will be listened to and issues like income inequality and environmental degradation will be seriously addressed. New challenges and opportunities abound. But the same litmus test for the success of any government remains. Is it able to deliver sustainable economic growth, the fruits of which are shared by most of the population? Can its people enjoy steady improvements in their standard of living, and thereby foster a stable, cohesive and harmonious society? Maintaining and further enhancing competitiveness of the economy is a crucial element for achieving such success. This subject constitutes a central feature of ACI's research activities.

Professor Wang Gungwu
Chairman, Governing Board
Lee Kuan Yew School of Public Policy
National University of Singapore

Preface

The biggest phenomenon of our time is the great convergence between the Third World and the First World. In the past, a very small percentage of the world's population enjoyed the comfortable living standards of Western middle classes. In 2010, about 500 million Asians enjoyed middle-class living standards. Between 2010 and 2020, the number of Asians enjoying middle-class living standards will increase 3.5 times, from 500 million to 1.75 billion. This is a dramatic increase. Equally remarkably, the National Intelligence Council (NIC) of the USA has predicted that global poverty will disappear by 2030. The human condition will improve significantly.

China has contributed to this mega-trend of global convergence. It has had stunning achievements in poverty reduction. Since initiating market reforms three-and-a-half decades ago, more than 600 million people have been rescued from absolute poverty in China.

The reason these Asian countries, and China in particular, are now succeeding in this way is because they have finally begun to understand, absorb and implement important reforms: free-market economics; mastery of science and technology; meritocracy; a culture of pragmatism; a culture of peace; the rule of law; plus, of course, rising education levels. One of the most important elements among these ingredients that constitute the recipe for success is the notion of competitiveness. The success of any market economy is dependent on the dynamism of its competitive forces. Meritocracy is also key for identifying, nurturing and utilising the talented members of the society through fair and equitable competition. The rule of law enables such fair competition, which in turn ensures the proper functioning of a market economy.

As such it is difficult to overstate the importance of the messages conveyed in this book, which is the third instalment of research on Greater China, conducted by the Asia Competitiveness Institute (ACI) at the Lee Kuan Yew School of Public Policy (LKYSPP), National University of Singapore (NUS). It also complements ACI's concurrent studies on the competitiveness of other major Asian economies including India, Indonesia and the member countries of the Association of Southeast Asian Nations (ASEAN).

Drawing from a rich database of relevant indicators built and managed by ACI and its uniquely crafted methodology to analyse competitiveness at sub-national levels, ACI has continued to make significant contributions to the current state of knowledge in this field. These studies also have great practical bearings. They help us understand the catalytic role played by competitiveness in the unfolding economic development of China. This is not only significant for China. It also has huge implications for Asia and the world.

After more than three decades of rapid and unprecedented growth, it is natural that problems, structural or cyclical in nature, would surface in China. Among them, the issue of regional imbalance in socioeconomic advancement is a prominent challenge. The timely publication of this book will offer insight for policy makers to help address concerns associated with regional development disparity.

The results of this competitiveness analysis on the 34 economies of Greater China were presented and the accompanying public policy recommendations discussed during ACI's November 2014 Annual Conference. The discussion revealed new insights on how the economies could seize the opportunity to further boost development through competent, effective and open-minded management. At a subsequent Review Seminar held on 28 August 2015 at the LKYSPP, an engaged audience, consisting of policy planners and managers, as well as experts discussed the findings on relative competitiveness performance of the 34 economies and five regions of Greater China. We believe that the facts and analysis presented in this book could usefully inform Greater China policy makers, business leaders, intellectuals and other stakeholders in Asia and beyond.

I am therefore proud to present this remarkable volume on competitiveness in the hope that it will help the Greater China economies and regions to further realise their potential. A stable and prosperous China will benefit Asia and the world. Both ACI and LKYSPP are happy to support this noble goal.

Professor Kishore Mahbubani
Dean
Lee Kuan Yew School of Public Policy
National University of Singapore

Acknowledgements

We from the Asia Competitiveness Institute (ACI) at Lee Kuan Yew School of Public Policy (LKYSPP), National University of Singapore (NUS) have been tracking the competitiveness of 34 Greater China economies, amongst many other Asian economies, since 2000. Our work includes published volumes in 2013 and 2014 on Competitiveness Analysis and Development Strategies for 33 Indonesian Provinces; Annual Analysis of Competitiveness, Simulation Studies and Development Perspective for 35 States and Federal Territories of India: 2000–2011; Annual Analysis of Competitiveness, Development Strategies and Public Policies on ASEAN-10: 2000–2011; and Annual Analysis of Competitiveness, Simulation Studies and Development Perspective for 34 Greater China economies. This volume provides an update by extending the study on Greater China economies to include 2000–2012.

In all of these publications, ACI makes the distinctive contribution to the literature by focusing on competitiveness at sub-national levels, over and above traditional national level analysis. In fact, ACI also drills down to cities as competitive, liveable and intelligent as well. It well recognised that it is cities which attract talents, capital and business interests to drive policies and strategies at all levels of government in commensurate. Equally important, ACI works on evidence-based analysis supported by quantitative methodologies as both scientifically and artfully presented to be useful to policy-makers.

In this book on "2015 Estimation of Agricultural Productivity and Annual Update of Competitiveness Analysis for Greater China: Optimising Agricultural Productivity and Promoting Innovation Driven Growth", we use the same unique methodology and tools as in our other studies of competitiveness. Over time, the methodology and indicators are continuously being improved, and indicators are regularly being reviewed and updated at the economy level. Taken together, ACI thus offers both macro economy level assessment and indicators at micro level by sectors for competitiveness evaluation. Effective coordination across-the-board is ensured in the context of political economy.

After decades of unbalanced growth in different regions of Greater China resulting in huge gaps in economic prosperity level and living standards, regional development highlights the urgent need to spread growth centres across Greater China. We have thus expanded the competitiveness analysis of Greater China to cover not only a longer temporal dimension, but also regional competitiveness analysis.

Following the academic tradition, prior to any book publication, ACI invites comments and suggestions as peer reviews from experts and practitioners in the field. Drawing from a Review Seminar held in LKYSPP, NUS on 28th August 2015 before the book is published, ACI received tremendous support from various experts of 34 Greater China economies who came from the public and private business sectors. We wish to record the effort and time taken by the reviewers in evaluating the chapters and their constructive comments are incorporated accordingly. We would therefore like to thank the following reviewers:

Professor Zhang Shucun
Dean, Shandong Academy of Social Sciences, People's Republic of China

Professor Chi Fulin
President, China Institute for Reform and Development and Member of the Chinese People's Political Consultative Conference, People's Republic of China

Professor Chang Xiuze
Professor, Academy of Macroeconomic Research, National Development and Reform Commission, People's Republic of China

Dr He Dongni
Vice President, China Institute for Reform and Development, People's Republic of China

Dr Ni Pengfei
Director, Centre for City and Competitiveness, Chinese Academy of Social Sciences, People's Republic of China

Dr Zhang Qingjin
Director, Institute of Rural Development, Shandong Academy of Social Sciences, People's Republic of China

We place on record our appreciation for the encouragement of Professor Kishore Mahbubani (Dean), Professor Kanti Prasad Bajpai (Vice Dean, Research), Associate Professor Kenneth Paul Tan (Vice Dean, Academic Affairs), Mr Donald Low (Associate Dean, Research and Executive Education) and other colleagues at LKYSPP in making this effort possible. This book would not be possible without the dedicated assistance of the research and administrative staff from ACI. They have been working hard throughout in organising the Review Seminar and overall rolling-out of the research project. Thus, we note with great appreciation the high team spirit of the ACI staff, including Blake Harley Berger, Brian Riady, Choo Yan Jia Nicholas, Evan Tan Beng Kai, Grace Aw Ee Ling, Ham Oi Mun, Harminder Chyle Rajan, Lian Xiao, Linda Low, Liu Dantong, Liu Muyan, Mayu Oba, Mulya Amri, Nie Tongxin, Nurina Merdikawati, Nurliyana Binte Yusoff,

Qin Jiao, Ramkishen S. Rajan, Rao Vittal Kartik, Ririn Radiawati Kusuma, Rizky Novrianto, Sangiita Yoong Wei Cher, Sujata Kaur, Suo Haoran, Tracy Cai Jiao, Vincent Kwan Wen Seng, Yan Yanying, Yap Xin Yi, Ye Ye Denise, Yip Fung Yi Estelle, Zhang Yuchen, Zhao Ziyi and Zhu Na.

Tan Khee Giap, Yuan Randong and Xie Teleixi
Asia Competitiveness Institute, Lee Kuan Yew School of Public Policy
National University of Singapore

About the Authors

Tan Khee Giap is a Co-Director of the Asia Competitiveness Institute (ACI) and Associate Professor at the Lee Kuan Yew School of Public Policy, National University of Singapore. He is also the Chairman of the Singapore National Committee for Pacific Economic Cooperation. Upon graduating with a PhD from University of East Anglia, England, in 1987 under the Overseas Research Scheme awarded by the Committee of Vice-Chancellors and Principals of the Universities of the United Kingdom, he joined the banking sector as a treasury manager. Thereafter, he taught at the Department of Economics and Statistics, National University of Singapore, 1990–1993. Dr Tan joined Nanyang Technological University in 1993 and was Associate Dean, Graduate Studies Office, 2007–2009. Dr Tan has consulted extensively with the various government ministries, statutory boards and government-linked companies of the Singapore government. He has also served as a consultant to international agencies such as the Asian Development Bank, Asian Development Bank Institute, United Nations Industrial Development Group, World Gold Council, ASEAN Secretariat, Central Policy Unit of Hong Kong, Kerzner International, Las Vegas Sands and Marina Bay Sands. Dr Tan is the lead author for twenty books, serves as journal editors and published widely in international refereed journals. His current research interests include econometric forecasting, cost of living index, global liveable cities index, ease of doing business index and competitiveness analysis on sub-national economies of China, India, Indonesia and ASEAN. Dr Tan was Deputy President of the Singapore Economic Society, 2004. He served in the 2002 Economic Review Committee (ERC), served as the Deputy Chairman of the IPS Forum for Economic Restructuring (IFER) in 2003 and served as a member of the Resource Panel of the Government Parliamentary Committee for Transport and Government Parliamentary Committee for Finance and Trade since 2007. Dr Tan has extensively advised and guided multinational corporations leading to public listings, especially those companies from Mainland China and Taiwan. He is also currently an Independent Director of the publicly listed Artivision, BreadTalk Group, Boustead Projects and TEE Land.

Yuan Randong is a Research Associate at the Asia Competitiveness Institute at the Lee Kuan Yew School of Public Policy, National University of Singapore. He obtained a joint degree from the National University of Singapore and Australian National University in Economics and Actuarial Studies with first class honours in 2012. At Asia Competitiveness Institute, he focuses on data analysis and competitiveness studies of the Greater China economies. Yuan Randong's research interests include financial economics, econometric modelling, and economic development and competitiveness of the Greater China economies. As a co-author, he published an editorial opinion on public policy in Singapore, a journal paper in *International Journal of Chinese Culture and Management*, and a book "Annual Analysis of Competitiveness, Simulation Studies and Development Perspective for 34 Greater China Economies".

Xie Teleixi is a Research Assistant at the Asia Competitiveness Institute at the Lee Kuan Yew School of Public Policy, National University of Singapore. Teleixi graduated with a first class honours bachelor's degree from the National University of Singapore in Economics in 2015 and was awarded the Dean's list twice during her studies. At Asia Competitiveness Institute, her work focuses on data analysis and competitiveness studies of the Greater China economies. She is also the coordinator for the project on the city-level competitiveness study for Shandong province in China. She has co-authored an editorial opinion on public policy in Singapore, a journal paper in the *Journal of Economics and Development Studies* (under review), and one book on "2015 Annual Competitiveness Analysis, Agricultural Productivity and Development Vision for ASEAN-10" (forthcoming). Teleixi's research interests include quantitative data analysis and modelling, labour economics and competitiveness analysis of Greater China economies.

List of Economies

	Name of Economies in English	Name of Economies in Chinese
1	Anhui	安徽
2	Beijing	北京
3	Chongqing	重庆
4	Fujian	福建
5	Gansu	甘肃
6	Guangdong	广东
7	Guangxi	广西
8	Guizhou	贵州
9	Hainan	海南
10	Hebei	河北
11	Heilongjiang	黑龙江
12	Henan	河南
13	Hong Kong	香港
14	Hubei	湖北
15	Hunan	湖南
16	Inner Mongolia	内蒙古
17	Jiangsu	江苏
18	Jiangxi	江西
19	Jilin	吉林
20	Liaoning	辽宁

(*Continued*)

(*Continued*)

	Name of Economies in English	Name of Economies in Chinese
21	Macau	澳门
22	Ningxia	宁夏
23	Qinghai	青海
24	Shaanxi	陕西
25	Shandong	山东
26	Shanghai	上海
27	Shanxi	山西
28	Sichuan	四川
29	Taiwan	台湾
30	Tianjin	天津
31	Tibet	西藏
32	Xinjiang	新疆
33	Yunnan	云南
34	Zhejiang	浙江

List of Abbreviations

ACI	Asia Competitiveness Institute
AIIB	Asian Infrastructure Investment Bank
APEC	Asia-Pacific Economic Cooperation
ASEAN	Association of Southeast Asian Nations
CPI	Consumer Price Index
DEA	Data Envelopment Analysis
EU	European Union
FBMC	Financial, Businesses and Manpower Conditions
FDI	Foreign Direct Investment
GCR	Global Competitiveness Report
GDP	Gross Domestic Product
GFC	Global Financial Crisis
GIS	Government and Institutional Setting
GNI	Gross National Income
GRDP	Gross Regional Domestic Product
ICT	Information and Communications Technology
IMD	International Institute of Management Development
IMF	International Monetary Fund
LKYSPP	Lee Kuan Yew School of Public Policy
MS	Macroeconomic Stability
NBS	National Bureau of Statistics
NUS	National University of Singapore
OBOR	One Belt One Road
QLID	Quality of Life and Infrastructure Development

R&D	Research and Development
SME	Small and Medium-sized Enterprise
TC	Technological Change
TEC	Technical Efficiency Change
TFP	Total Factor Productivity
TFPC	Total Factor Productivity Change
US	United States
USD	United States Dollar
WCY	World Competitiveness Yearbook
WTO	World Trade Organisation

List of Tables and Figures

FIGURES

Chapter 1

Chapter 2

Chapter 3

Chapter 4

TABLES

Chapter 2

Chapter 3

Chapter 4

Chapter 1

Organisation, Structure and Methodology

1.1 WHY THIS BOOK AND FOR WHOM

In the tradition of the Asia Competitiveness Institute (ACI) at the Lee Kuan Yew School of Public Policy (LKYSPP), National University of Singapore (NUS), our work and publications make a distinctive contribution to the literature, firstly by focusing on competitiveness analysis and ranking at sub-national levels, over and above traditional analysis at national levels.

Secondly, it is in this spirit that we at the ACI have crafted a comprehensive methodology (the details of which are covered in the next section of this chapter) to assess annual competitiveness at the sub-national level. This methodology is employed here in this book, and in ACI's previous publications reporting research findings on Greater China, Indonesia, India and ASEAN countries (Tan *et al.* 2013; Tan *et al.* 2013; Tan *et al.* 2013a, 2013b; Tan *et al.* 2014).

Since the inaugural volume of the *Annual Analysis of Competitiveness, Simulation Studies and Development Perspective for 34 Greater China Economies*: 2000–2010 (Tan *et al.* 2014), the comprehensive competitiveness index developed by ACI has been used to rank the 34 Greater China economies in various dimensions. As ACI aims to provide an update on findings of all studies, this updated book thus includes data from 2000 to 2012 for 34 Greater China Economies.

ACI's evidence-based analysis, which is supported by quantitative methodologies, offers great value for policymakers, academicians, experts, practitioners and players in business and non-governmental organisations, all of whom are among ACI's target audience. Following the academic tradition of multidisciplinary and cross-border dialogues among stakeholders, ACI holds Review Seminars prior to any book publication. As noted in the Acknowledgements, the Review Seminar on Provincial and Regional Competitiveness Analysis for Greater China was held in LKYSPP, NUS on 28th August 2015.

This book, comprising five chapters, has this introductory chapter as a road map for readers. Chapter 2 details the 2015 Annual Update on Competitiveness Ranking and Simulation Studies for 34 Greater China Economies. Chapter 3 features the results of the 2015 Regional Competitiveness Analysis on the five regions of Greater China, namely Eastern Coastal Area of China, Central China, Western China, Northeastern China and Hong Kong, Macau and Taiwan.

In Chapter 4, we report the results of the analysis on the Agricultural Productivity in Greater China Economies, which focuses on studying the provincial agricultural Total Factor Productivity (TFP) in mainland China. Chapter 5 concludes our views on China's future growth prospect and the way forward towards an innovation-driven economy for Greater China. Potential research agenda by ACI is also highlighted in this chapter. Three appendices feature ACI's list of indicators for provincial level economies, the algorithm for competitiveness analysis, and ACI's regional competitiveness list of indicators with notes on data aggregation method.

The rest of Chapter 1 is organised as follows. Section 1.2 details the methodology and data used to derive the competitiveness rankings. It also includes the approach adopted by ACI to conduct policy-induced simulation analysis on the 34 Greater China economies, to allow for improvements in competitiveness on various dimensions. This makes the contribution much more constructive than other indices that merely report rankings.

1.2 METHODOLOGY AND DATA SOURCES

1.2.1 Literature Review on Competitiveness

The notion of regional or *in situ* competitiveness has been explored in various ways. Initially, in the 1980s, the term "competitiveness" was considered as part of the private sector domain, especially in relation to firms performance (Porter 1980). Over time, the discourse expanded onto the domain of macroeconomics, as shown by the establishment of "Council on Competitiveness" or similar institution in various advanced industrialised countries. Some scholars acknowledge that it is primarily firms that compete, not nations, regions or places (Krugman 1994). However, the performance of firms depends on various factors that are embedded in the firm's location. These include quality of labour and infrastructure, governance, costs of doing business, as well as the performance of competing and complementary firms around the area (Bristow 2010; Camagni 2002; Kitson, Martin and Tyler 2004).

Two of the most highly cited cross-country competitiveness studies at the moment are the World Competitiveness Yearbook (WCY) and the Global Competitiveness Report (GCR). The WCY defines competitiveness as composed of four factors: (1) economic performance, (2) government efficiency, (3) business efficiency and (4) infrastructure. Each factor is further made up of five criteria, so that there are 20 criteria in total. The WCY uses equal weights such that each criterion contributes 5% weight, and each factor contributes 25% weight to the overall competitiveness index (World Competitiveness Center 2014). Most of the data used are statistical indicators sourced from international, regional and national organisations. Substantial perception data sourced from executive surveys also supplement the statistical data.

The GCR breaks down its Global Competitiveness Index into three sub-indices: (1) basic requirements, (2) efficiency enhancers and (3) innovation and sophistication factors (World Economic Forum 2014). Each of the three sub-indices has a different number of pillars. The basic requirements sub-index has four pillars: institutions, infrastructure, macroeconomic environment and health and primary education. The efficiency enhancers sub-index is composed of six pillars: higher education and training, goods market efficiency, labour market efficiency, financial market development, technological readiness and market size. Lastly, the innovation and sophistication factors sub-index consists of two pillars: business sophistication and innovation. Most of the data that make up these pillars and sub-indices come from executive opinion surveys.

1.2.2 Four Environments and 11 Sub-Environments

Our approach to competitiveness is comprehensive because it takes into account different factors that collectively shape the ability of an economy to achieve substantial and inclusive economic development over a sustainable period of time.

Guiding this overarching context, ACI has defined competitiveness according to four environments: (1) Macroeconomic Stability, (2) Government and Institutional Setting, (3) Financial, Businesses and Manpower Conditions and (4) Quality of Life and Infrastructure Development. These four environments have the same impact on Overall Competitiveness, wherein each environment counts for 25% of Overall Competitiveness score (see Figure 1.1).

ACI's competitiveness framework also uses a nested approach, creating two to three sub-environments under each of the four environments. There are a total of 11 sub-environments for the 34 Greater China economies. Each sub-environment within each environment is given equal influence as the others under the same environment.

ACI uses a simple averaging mechanism to integrate all sub-environments (each is given the same weight) into their respective environment. Thereafter, all environments are combined

Figure 1.1: ACI's Competitiveness Framework

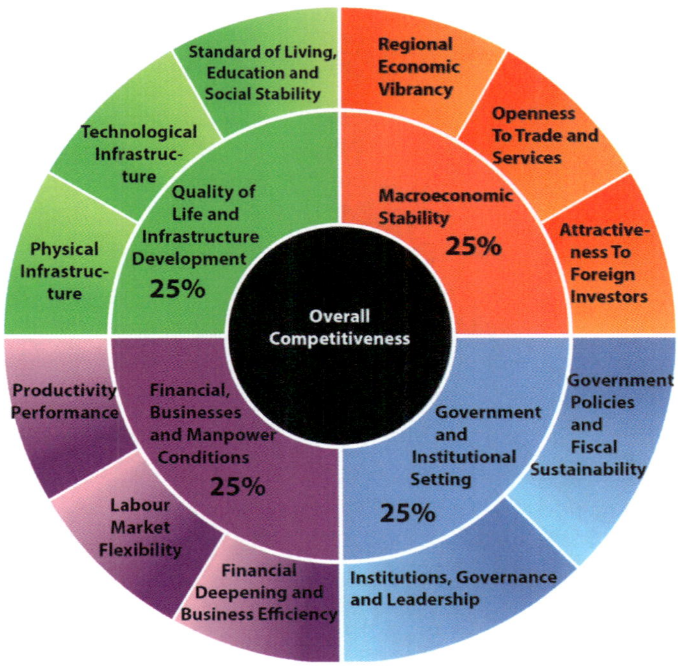

Source: Asia Competitiveness Institute

into an Overall Competitiveness ranking. While assigning different weight to each environment, sub-environment and indicator based on their importance may be more appropriate, the implementations are controversial. Therefore, to achieve a balanced view for each factor of Overall Competitiveness, equal weights are assigned.

The ACI competitiveness framework, with its four environments, is a comprehensive and systematic way of examining competitiveness. Two of these environments are directly linked to the economy. Macroeconomic Stability expresses aggregate economic activity and includes three sub-environments: (1) Regional Economic Vibrancy, (2) Openness to Trade and Services and (3) Attractiveness to Foreign Investors. Regional Economic Vibrancy includes data on an economy's Gross Regional Domestic Product (GRDP), value-addedness of the three economic sectors, Consumer Price Index (CPI), tourism and investment. Openness to Trade and Services goes further to examine how international an economy is, taking into consideration the position of its merchandise on the international market as well as the performance of trade and tourism. Attractiveness to Foreign Investors includes the state of foreign direct investment and foreign-funded companies that are active in a particular economy.

Financial, Businesses and Manpower Conditions mainly looks at the microeconomic state of the economy, examining its financial strength and understanding the factors that can affect a company's productivity, especially labour force. This environment includes three sub-environments: (1) Financial Deepening and Business Efficiency, (2) Labour Market Flexibility and (3) Productivity Performance. Financial Deepening and Business Efficiency reflects the state of financing and investment in economic activities in an economy. Labour Market Flexibility includes data on employment and wages. Productivity Performance includes the value-added productivity per worker for each of the three economic sectors, investment in fixed asset and overall labour productivity.

The other two environments are identified as Government and Institutional Setting and Quality of Life and Infrastructure Development. Government and Institutional Setting includes the fiscal state of government and economic structure of a particular economy. This environment includes two sub-environments: (1) Government Policies and Fiscal Sustainability and (2) Institutions, Governance and Leadership. Government Policies and Fiscal Sustainability examines the expenses, revenue and tax collection for each economy. Institutions, Governance and Leadership looks at the corporate structure of an economy to measure how market-oriented it is. Data used include that for domestically funded enterprises, state-owned enterprises, collective-owned enterprises, limited liability corporations, joint-stock companies, private enterprises and foreign funded enterprises.

Finally, Quality of Life and Infrastructure Development takes an overall view of basic infrastructure and public services as well as the overall quality of life. The three sub-environments under this environment include: (1) Physical Infrastructure, (2) Technological Infrastructure and (3) Standard of Living, Education and Social Stability. Physical Infrastructure mainly refers to the level of development in an economy in terms of transportation (railways, highways, navigable inland waterways, and civil aviation routes). Among the 34 Greater China economies, several economies lag further behind others due to geographical factors. For example, neither Beijing nor Tibet has navigable inland waterways. For such economies, data has been assigned as zero. In addition to transportation, Physical Infrastructure also includes the production levels of natural resources (for example, coal, oil and iron) of an economy. Similar to transportation data, some of these values may also be zero depending on the kind of natural resources each economy has. For instance, Hainan and Tibet produce very little coal. We have done our best to collect a range of transportation and natural resource data to reflect as accurately, the advantages and disadvantages of each economy. Technological Infrastructure generally includes information and communications technology (ICT) such as personal computers, televisions, fixed telephones, mobile telephones, internet access and patents. Meanwhile, healthcare, education, population structure and the state of the environment are all included in the Standard of Living, Education and Social Stability sub-environment.

1.2.3 Indicators

There are a total of 102 indicators used in the competitiveness ranking of the 34 Greater China economies. The availability of data is greater for indicators in certain environments and sub-environments. See Appendix 1 for a complete list of indicators. To a certain extent, having a higher number of indicators ensures more robust interpretation of the aggregated sub-environment score. However, the uneven distribution of indicators across sub-environments arising from data availability is not an issue. Scores of indicators will average out regardless of the number of indicators in a sub-environment. This provides flexibility in deciding whether to add or reduce the number of indicators in the future, as long as the overall framework of four environments and 11 sub-environments remains unchanged.

1.2.4 Data Sources and Aggregation of Economies Data for Indicators for Regions

The 102 indicators used in this research are derived from a number of sources. First among these are official government statistics from the National Bureau of Statistics of People's Republic of China (NBS), various statistical bureaus at the provincial level, the Hong Kong Census and Statistics Department, the Macau Statistics and Census Service, the National Statistics of the Republic of China (Taiwan), the World Bank and the United Nations.

The way to construct the competitiveness ranking by region follows the same methodology used for the competitiveness ranking by economy, with appropriate dataset for indicators at the regional level. As each region consists of a few economies, data aggregation is adopted to extend the ranking analysis from sub-national level to regional level.

Many of the indicators used for the sub-national (i.e. economy) ranking can be directly aggregated to get the indicator values for a region by summing across all the constituent economies in that region, as long as those indicators are based on absolute quantities, such as GRDP and Population. For indicators that are defined as ratios, we can compute them according to their definition once we have aggregated the absolute quantities upon which they are defined (i.e. aggregating the denominator and numerator first).

As an illustration, below are four examples for data aggregation to derive the regional indicator values. The first example is for indicator GRDP. Since it is based on absolute quantity, we can get the regional value by summing across all the economies within this region. The second example is for indicator GRDP Growth. Once we have computed the GRDP for each region across the years according to the way shown by Example 1, it is straightforward to calculate GRDP Growth by region according to its definition. The third example is for Gross Capital Formation. According to its definition, it is calculated as the ratio between the gross capital formation in the region and the GRDP for that region. Since both the numerator and denominator are absolute quantities, we aggregate them across all the economies within the region to get the regional values, from which we compute the ratio, hence arriving at the regional value for indicator Gross Capital Formation. Lastly, the fourth example is for Inflation Rate (CPI). Since the data is in the form of a percentage that cannot be easily disaggregated into numerator and denominator, the regional value for this indicator is obtained by taking the weighted average of economy-level data. In this case, we take the weighted average based on population size. In other instances, we can also take the weighted average based on land size, depending on the nature of the indicator.

***Example 1*: GRDP; Unit: 100 million yuan, Constant Prices at 2000**
For Region A with n economies at time t,

$$\text{GRDP}_{A,t} = \sum_{i=1}^{n} \text{GRDP}_{i,t}.$$

***Example 2*: GRDP Growth; Unit: Percentage Change Per Annum**
For Region A with n economies at time t,

$$\text{GRDP Growth}_{A,t} = \frac{\text{GRDP}_{A,t} - \text{GRDP}_{A,t-1}}{\text{GRDP}_{A,t-1}}.$$

***Example 3*: Gross Capital Formation; Unit: Percentage of GRDP**
For Region A with n economies at time t,

$$\text{Gross Capital Formation}_{A,t} = \frac{\sum_{i=1}^{n} \text{Gross Capital Formation}_{i,t}}{\sum_{i=1}^{n} \text{GRDP}_{i,t}}.$$

***Example 4*: Inflation Rate (CPI); Unit: Percentage Change per Annum**
For Region A with n economies at time t,

$$\text{Inflation Rate}_{A,t} = \frac{\sum_{i=1}^{n} \left(\text{Population}_{i,t} \times \text{Inflation Rate}_{i,t}\right)}{\sum_{i=1}^{n} \text{Population}_{i,t}}.$$

1.2.5 Standardised Score

The following will answer the question of how we processed our data and put them into a format that could be systematically analysed. The 102 indicators used in this research have different definitions and use different units of measure. For instance, the GRDP is measured in 100 million Renminbi while the population is measured in the number of people. In this section, we will explore how to resolve the issues caused by differences in definitions and units of measure for each indicator.

ACI uses a "standardised score" statistical method. The standardised score measures the performance difference between a specific economy and the average economy. There is no need to consider the use of different units of measure for different indicators because the standardised score uses no unit of measure. What is measured is the relative performance of various economies. In statistical terms, the standardised score measures how many standard deviations each economy is away from the average economy. See Appendix 2 for a technical explanation.

If the standardised score for an economy is zero, this means that the economy is an average performer in the 34 Greater China economies. If the standardised score is negative, this indicates that the economy is performing more poorly than others, while if it is positive, this indicates that it is performing better than others. The further the score is from zero, the further is the performance of the economy from the average of 34 Greater China economies. In this sense, a large negative score indicates very weak competitiveness, while a large positive score indicates very strong competitiveness.

The standardised scores for each indicator are then aggregated at the sub-environment level, and then re-aggregated at the environment level, and finally aggregated again at the overall level. This enables us to examine the performance of each of the 34 Greater China economies at various levels, from the Overall Competitiveness to specific indicators.

1.2.6 What-if Competitiveness Simulation Analysis

Competitiveness ranking is like a beauty pageant, because it only shows a list of those that perform well and those that face challenges without providing constructive recommendations that can help improve competitiveness. ACI takes one step further by tackling the "so what" question. What impact do competitiveness rankings have on the policy of a specific economy? The data used by ACI enable us to examine the indicators, sub-environments and environments of each economy in a targeted way. By analysing this data, not only we are able to provide a competitiveness ranking, we are also able to assess which indicators are weak and which are strong. This then allows us to provide policy suggestions that are specific to each economy.

The ACI's What-if Competitiveness Simulation Analysis looks at how to improve the top 20% weakest indicators for an economy, and provides a re-calculation of the standardised score based on this improvement. First, we put all of the indicators for a specific economy in order from lowest to highest based on their score. We are then able to identify the top 20 weakest indicators for that economy (20% of the total 102 indicators is 20).

Under most situations, we raise the scores of the top 20 weakest indicators up to the average scores for all 34 Greater China economies. This means that these 20 indicators will be changed from negative to zero. There are special cases for some economies in which some of the top 20% weakest indicators are in fact above the average scores. For these specific indicators, the score remains at the original value so that the score of the economy is not discounted. After increasing the standardised score for the top 20% weakest indicators, we recalculate rankings with the assumption that the scores for other economies remain unchanged. Only one economy is analysed during each simulation analysis. After completing simulation analysis of each of the 34 Greater China economies, each economy is given a new standardised score based on the simulated model. This can help us answer the question of how a particular economy's competitiveness ranking would change if its weakest indicators improve while other economies' scores stay constant.

REFERENCES

Bristow, G. 2010. *Critical Reflections on Regional Competitiveness: Theory, Policy, Practice (Routledge Studies in Human Geograph)*. Abingdon: Routledge.

Camagni, R. 2002. On the Concept of Territorial Competitiveness: Sound or Misleading? *Urban Studies* 39(13): 2395–2411.

Kitson, M., R. Martin and P. Tyler, 2004. Regional Competitiveness: An Elusive yet Key Concept? *Regional Studies* 38(9): 991–999.

Krugman, P. 1994. Competitiveness: A Dangerous Obsession. *Foreign Affairs* 73(2): 28–44.

Porter, M. E. 1980. *Competitive Strategy: Techniques for Analyzing Industries and Competitors*. New York: Free Press.

Tan, K.G., L. Low, K.Y. Tan and A.L. Lim, 2013. *Annual Analysis of Competitiveness, Development Strategies and Public Policies on ASEAN-10: 2000–2010*. Singapore: Pearson.

Tan, K.G., L. Low, K.Y. Tan and K. Rao, 2013. *Annual Analysis of Competitiveness, Simulation Studies and Development Perspective for 35 States and Federal Territories of India: 2000–2010*. Singapore: World Scientific.

Tan, K.G., M. Amri, L. Low and K.Y. Tan 2013a. *Analisis Daya Saing dan Strategi Pembangunan untuk 33 Provinsi Indonesia (Competitiveness Analysis and Development Strategies for 33 Indonesian Provinces) (in Bahasa Indonesia)*. Singapore: World Scientific.

Tan, K.G., M. Amri, L. Low and K.Y. Tan 2013b. *Competitiveness Analysis and Development Strategies for 33 Indonesian Provinces*. Singapore: World Scientific.

Tan, K.G., R. Yuan, S.W.C. Yoong and M. Yang 2014. *Annual Analysis of Competitiveness, Simulation Studies and Development Perspective for 34 Greater China Economies: 2000–2010*. Singapore: World Scientific.

World Competitiveness Center, I. for M. D. (IMD) 2014. *IMD World Competitiveness Yearbook*. Lausanne: IMD.

World Economic Forum. 2014. *Global Competitiveness Report 2013–2014*, (ed. K. Schwab). Geneva: World Economic Forum.

Chapter 2

2015 Annual Update on Competitiveness Ranking and Simulation Studies for 34 Greater China Economies

2.1 STRUCTURAL CHANGE AND FURTHER DEVELOPMENT: AN OVERVIEW OF GREATER CHINA

There would be little argument if one proclaims that the economic transformation in China thus far is a triumph in human development history. Figure 2.1 gives an illustration of China's phenomenal growth in the context of East Asia and compares it to the United States (US) and Japan. The real question lies in the future, as the medium or short-term prospect of China's growth and sustainability draw doubts among stakeholders and commentators at home and abroad.

At the crux of the dilemma China encounters in its endeavour for further development, is the syndrome of "excessive capacity", which has been identified as one of the fundamental reasons behind China's initiatives in founding the Asian Infrastructure Investment Bank (AIIB) and championing the course of international economic collaboration along the Silk Road Economic Belt and the 21st Century Maritime Silk Road, or One Belt, One Road (OBOR) (Dollar 2015; Li *et al.* 2015; Nataraj and Sekhani 2015).

At the current juncture as the top leaders of China have astutely identified the new kinds of parameters they face in continuing their reform efforts, they have also determined a novel combination of the challenges and opportunities facing Chinese economy, as defined in an apt, media-friendly term, the New Normal Era (Tan, Yuan, and Yoong 2015).

China's Thirteenth Five-Year Plan which has been rigorously mooted in multiple rounds of consultations with government counsellors, experts from various fields and representatives from the general public has been passed by the Central Committee of the Chinese Communist Party, and is expected to be fully released and come into force in March 2016. Many goals and reform initiatives are set in line with an overall plan to restructure the economy, from being heavily reliant on labour-intensive, low-skilled manufacturing, export and investment to becoming an innovation-driven economy that thrives on high-skilled manufacturing and value-added services and an invigorated level of domestic consumption. Although absolutely sensible and essential for

Figure 2.1: Gross Domestic Product of East Asia,[1] Japan, Mainland China and the United States from 1980 to 2014

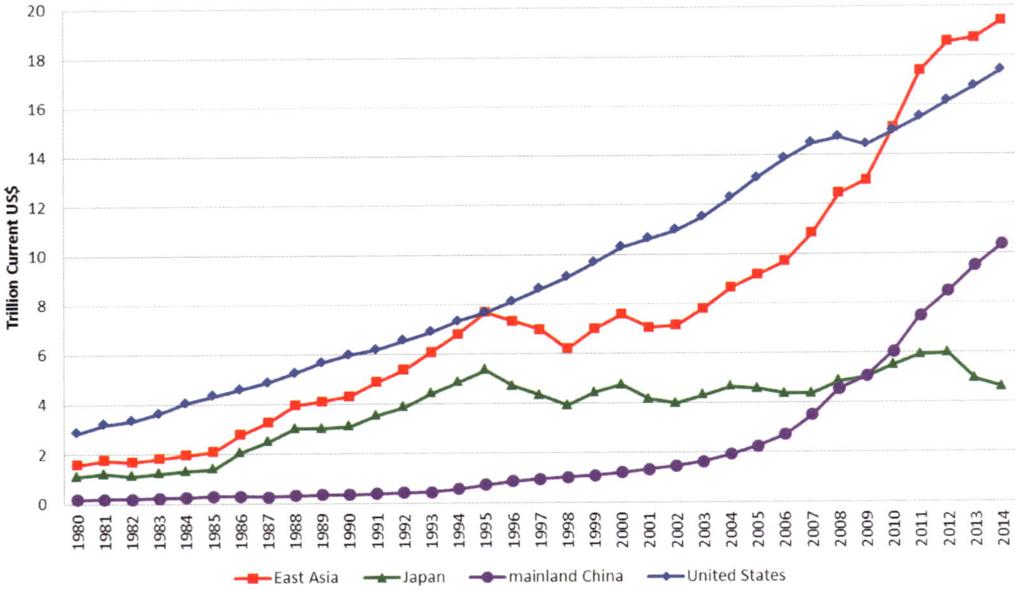

Source: Asia Competitiveness Institute (ACI) based on data obtained from World Development Indicators, World Bank

the sustainable growth of China in the medium to long term, such structural change in the economy remains a daunting task ahead, due to the malicious syndrome of "excessive capacity" China now suffers.

2.1.1 What Underpins "Excessive Capacity"

Capital goods face declining domestic demands with China's infrastructure and real estate development encountering a slowdown. The OBOR, AIIB and other initiatives by China to export its capital goods to other countries and efforts to capitalise on its ability in infrastructure development to create a win–win situation with its international partners, are great visions that receive positive responses from many countries. Nonetheless, these remain an uncertain glory as the complex and dynamic global economic outlook harbours risks that could unfavourably influence the implementation of such strategies, especially when combined with the ongoing sensitive geopolitical issues across the region.

Domestic demands for consumer goods are improving, epitomised by record breaking sales registered in the Single's Day in each of the last three years (i.e. Single's Day in China on 11th November 2013, 2014 and 2015) in Tmall and Taobao, the Chinese equivalent of Amazon and eBay. Revenues from these sales are several times larger than their American counterparts, at

[1] East Asia includes ASEAN-4 (Indonesia, Malaysia, Philippines and Thailand), Hong Kong, Japan, mainland China, Singapore, South Korea and Taiwan.

US$ 5.8 billion in 2013, US$ 9.3 billion in 2014, and over US$ 14.3 billion in 2015. Nonetheless, one should not be complacent as the increase in online shopping by Chinese consumers may not be an accurate barometer for overall consumption in the national economy. Indeed, we need to see over the longer term if the spike in sales in such days is not at the cost of too much erosion of sales in other nondescript days. Even if online sales have increased substantially in China, we should take a more comprehensive approach to include assessing other avenues of distribution network of consumer goods. One may hope to play a "win–win" game, but should always check if it is in fact a zero-sum one.

The old model of investment-driven growth can be considered as one of the root causes for excessive capacity, which induced over investment in selected few industries, at levels substantially higher than required for future growth in domestic and foreign demands.

Over-competition among provinces and regions, and undue intervention of some local governments (albeit with good intention to safeguard local economic growth) could have exacerbated the situation. This prompts the need for greater coordination efforts across provinces and regions in order to avoid excessive build-up of capacity in similar industries with repeated and often low quality investment projects. This also allows efforts to proactively entertain strategic merger and acquisition attempts by Chinese companies with the core competitiveness in its products and technology to compete profitably, domestically and abroad.

The chasm between rich and poor also serves to worsen the situation for China with excessive capacity. The majority of the Chinese residents, in rural or urban areas, have long before been elevated from abject poverty. Few suffer from starvation nowadays in China. But few are also rich enough to buy the high quality consumer goods, the increasing demand of which is a major impetus to drive a country from middle-income level to high-income level. This is so, as demands of well-to-do households exert a form of force that is similar to the Darwinian selection, stimulating firms to make upscale products with the right kind of quality.

Nevertheless, the upgraded demands of Chinese consumers for better quality products have a potential role to play in promoting the industrial upgrading and economic restructuring in general, as observed and articulated by Premier Li Keqiang in the State Council executive meetings in November 2015 (Li 2015). The policymakers in the Central Government are aware of the fact that China known widely as 'factory of the world', is currently unable to meet the demands of the relatively rich consumers in China for high-end products (Yu 2015). Hence, China is the largest luxury goods importer of the world.[2] The lack of a sizeable upper-middle class may also explain why the sophistication in products design and quality improvement has not gained traction in many Chinese companies.

2.1.2 Productivity, Innovation and Human Capital: The Key Factors for Becoming a High-Income Country

Market clearance seems to be the right way for curing the malady of excessive capacity. It may be a painful process that is bound to encounter fierce resistance from vested interest groups for

[2]In terms of the nationality of the buyers, China has become the largest luxury market in the world since 2013, and it is still growing fast, according to *The Economist* (2013).

whom the best situation is the status quo. But any delay in economic reform and giving the market more voice in areas of the economy previously intervened heavily by the government will only delay the impacts of excessive capacity. Such indecision in action could also mean that the best window of opportunity to re-establish a structurally more balanced and sustainable model of development is missed.

Productivity, innovation and human capital are also the determinants of success of the structural change for the Chinese economy. Measured properly, they can function as informative indicators for the diagnostics and prognostics of an economy facing similar problems and striving towards a high income, developed country.

The productivity of the overall economy is the primary factor to determine whether a country is developing or has developed and what level it is at in the spectrum of national income per capita, a decisive measure of the average standard of living of its citizens. The agricultural productivity is of special importance to a still rapidly urbanising China and deserves a discussion on its own, as covered in Chapter 4.

The economy's ability to innovate depends significantly on the level of human capital it possesses. On this front, China has done well, in particular in two aspects that determine the quality and amount of the human capital. The first aspect is public health profile, a key factor which is often less regarded in the discourse of determinants of human capital but plays a critical role in deciding the level of human capital available to a country (Becker 2007).

To save the immiserating situation of extremely poor public health conditions that plagued China for decades since the turn of the 20th century, the first generation of the top leaders in China after 1949 made huge efforts to solve the issue of debilitating infectious diseases that were then epidemic in the country. Despite the very low income level for China then, the leaders mobilised a huge amount of resources and exercised great coordination efforts to drastically reduce the prevalence of such diseases that could have spread to a much wider portion of the population had such proactive measures did not take place (Liu 2004). As a salient example, in her speech delivered for the 2015 Nobel Prize in Physiology or Medicine, Tu Youyou expressed her appreciation for the great support by the Chinese political leaders and their contribution in fighting against infectious diseases in the 1960s, and paid her tribute by quoting Mao Zedong's hand written directive that "Chinese medicine and pharmacology are a great treasure-house, and we should explore them and raise them to a higher level".[3]

The result was manifest in very pronounced increase in the life expectancy of Chinese people in the first two to three decades after the founding of the People's Republic of China, greatly boosting the public health profile of the country, thereby adding many productive years of healthy lives for its people to live, work and enjoy.[4]

[3]The video recording of Tu Youyou's speech for 2015 Nobel Prize in Physiology or Medicine is available at http://www.nobelprize.org/nobel_prizes/medicine/laureates/2015/tu-lecture.html. It has been reported that answering the appeal by Mao Zedong to find cures for malaria led Tu Youyou to the Nobel Prize (Perlez 2015). The fact that the very first Chinese citizen awarded with the Nobel Prize for a science subject is a woman speaks volumes about the greatly improved gender equality in access to education and career opportunities in China since 1949 (Zhu 2015).

[4]Consequently, the average life span for the people increased dramatically in the 1960s in China. In only one decade, the life expectancy at birth increased by almost 20 years, from 43.5 years in 1960 to 62.9 years in 1970, according to World Bank data.

The other area that bodes well for China is its strength in education. Education is a form of mental or spiritual liberation and a type of investment that is widely accepted as the primary determinant of the quality and quantity of human capital (Friedman *et al.* 2011; Becker 1994). It is the necessary avenue through which a country becomes truly developed and via which its economy gains sophistication and derives its core competitiveness. The education system in China and its evolution since 1949 has helped to radically decrease the rate of illiteracy among its young and adult population.

The tertiary education reform which plays a small but critical part of reform and the opening up of China since the late 1970s have increased the number of college and university graduates every year from an insignificant number of less than 150,000 (as compared to China's population of close to one billion then) to 7.5 million in 2015, a rise of more than 50 folds in the number of annual graduates.[5] The accumulative result of such annual increase is prodigious, as it translates into a formidable force of 69 million graduates into the Chinese economy since 2000. The catalytic effect of such education reform in China will go a long way towards helping the process of the structural change currently required to elevate China from an economy that manufactures low-end products to an innovative trendsetter that prevails in the higher ends of the global value chain.

The rest of Chapter 2 is organised as follows. Section 2.2 discusses research findings which include the overall competitiveness ranking and the competitiveness rankings of four environments. Changes in the competitiveness of the 34 Greater China economies over time are also tracked. This is followed by the results of policy-induced simulation analysis that reveal potential improvements for each economy. Section 2.3 concludes by summarising salient results of the study.

2.2 RESEARCH FINDINGS

After describing how standardised scores are obtained (see Chapter 1, Section 1.2 on Methodology and Data Sources), we would now like to discuss the 2015 rankings for the 34 Greater China economies. All rankings, Overall Competitiveness ranking and the rankings for the four environments, are ordered based on the standardised scores of each economy. We will also include results of the What-if Competitiveness Simulation Analysis.

2.2.1 ACI 2015 Overall Competitiveness Ranking and Score

Table 2.1 shows the 2015 Overall Competitiveness ranking. From the spread, we can see that the maximum score for the 34 Greater China economies is 2.0367 and the minimum score is −1.6789. Guangdong is ranked top among the 34 economies with the highest score while Tibet is ranked #34.

[5]The number of annual graduates from Chinese universities and colleges in 2015 is reported in an article in *Guangming Daily*, published on 17th July 2015 (MyCOS 2015). The number of annual graduates from universities and colleges in China in the late 1970s to be less than 150,000 is deduced from the fact that in 1980 the number of annual graduates was 139,640 in China, according to *China Statistical Yearbook*.

The average performing economy (with a standardised score of 0) is somewhere between the #13-ranked Hebei (0.0595) and the #14-ranked Sichuan (–0.1626). As for the median, which is at the middle of a ranking, it is situated between #17-ranked Henan (–0.2842) and the #18-ranked Shaanxi (–0.3299). The median score for the 2015 Overall Competitiveness ranking is thus –0.3071.[6] Clearly, we can see that the average score (0) is slightly higher than the median score (–0.3071). The average value is positioned in the upper half of the ranking, above the centre of the list.

For the distribution of the Overall Competitiveness score, we found that there are several gaps in scores between economies. For example, there is a relatively large gap between the #8-ranked Shandong (0.9132) and the #9-ranked Liaoning (0.4251), their scores differ by 0.4881. There is also a considerable gap between the #33-ranked Gansu (–1.2443) and #34-ranked Tibet (–1.6789), where the difference in scores is 0.4346. The same goes for the #7-ranked Zhejiang (1.3244) and the #8-ranked Shandong (0.9132), whereby the gap in scores between them is 0.4112. These large gaps are sizeable despite being only one ranking apart.

Table 2.1: ACI 2015 Overall Competitiveness Ranking and Score[7]

Rank	Economy	Score
1	Guangdong	2.0367
2	Jiangsu	1.9901
3	Taiwan	1.7056
4	Beijing	1.6291
5	Hong Kong	1.6120
6	Shanghai	1.5189
7	Zhejiang	1.3244
8	Shandong	0.9132
9	Liaoning	0.4251
10	Macau	0.4167
11	Fujian	0.1578
12	Tianjin	0.0682
13	Hebei	0.0595
14	Sichuan	–0.1626
15	Hubei	–0.1833
16	Inner Mongolia	–0.2043
17	Henan	–0.2842

(Continued)

[6]Since there is an even number of economies in our ranking (i.e. 34 economies), the median is the mean of the two middle values. [(–0.2842) + (–0.3299)]/2 = –0.3071.

[7]Result shown in Table 2.1 is based on 2012 data, collated and analysed in 2015.

Table 2.1: *(Continued)*

Rank	Economy	Score
18	Shaanxi	–0.3299
19	Anhui	–0.3646
20	Xinjiang	–0.3816
21	Jilin	–0.4076
22	Heilongjiang	–0.4317
23	Hunan	–0.5517
24	Shanxi	–0.5588
25	Yunnan	–0.6348
26	Chongqing	–0.6536
27	Jiangxi	–0.7870
28	Hainan	–0.8450
29	Guangxi	–0.8529
30	Ningxia	–0.9870
31	Qinghai	–1.1515
32	Guizhou	–1.1619
33	Gansu	–1.2443
34	Tibet	–1.6789

Source: Asia Competitiveness Institute

We can also look at the rankings of 34 Greater China economies from a geographical and spatial perspective. In Figure 2.2, on the 2015 Overall Competitiveness ranking map, we have colour-coded each economy according to three categories. Green represents the top 10 economies, yellow, the middle 14 economies, and red, the 10 lowest ranking economies. We see that the green mainly covers the coastal areas in mainland China, except for three economies, Fujian, Hebei and Tianjin. Hong Kong, Macau and Taiwan are in green as well. This indicates that the economies in mainland China coastal area as well as Hong Kong, Macau and Taiwan are comparatively more competitive than the rest of the economies in Greater China.

On the other hand, economies located in the Central and Northeastern area of mainland China are largely yellow while economies in the western China tend to be red. This seems to be in line with the stepped development theory (Tan *et al.* 2014). However, a closer look at Figure 2.2 reveals that the green, yellow and red demarcations of economies do not exactly match the official regional classification as announced by National Bureau of Statistics (NBS).[8] Not only is the

[8]The four economic zones in mainland China as defined by the NBS of the People's Republic of China are Eastern Coastal Area in China, Central China, Western China and Northeastern China. The Eastern Coastal Area includes Beijing, Fujian, Guangdong, Hainan, Hebei, Jiangsu, Shandong, Shanghai, Tianjin and Zhejiang. Central China includes Anhui, Henan, Hubei, Hunan, Jiangxi and Shanxi. Western China includes Chongqing, Gansu, Guangxi, Guizhou, Inner Mongolia, Ningxia, Qinghai, Shaanxi, Sichuan, Tibet, Xinjiang and Yunnan.While Northeastern China

Figure 2.2: ACI 2015 Overall Competitiveness Ranking Map[9]

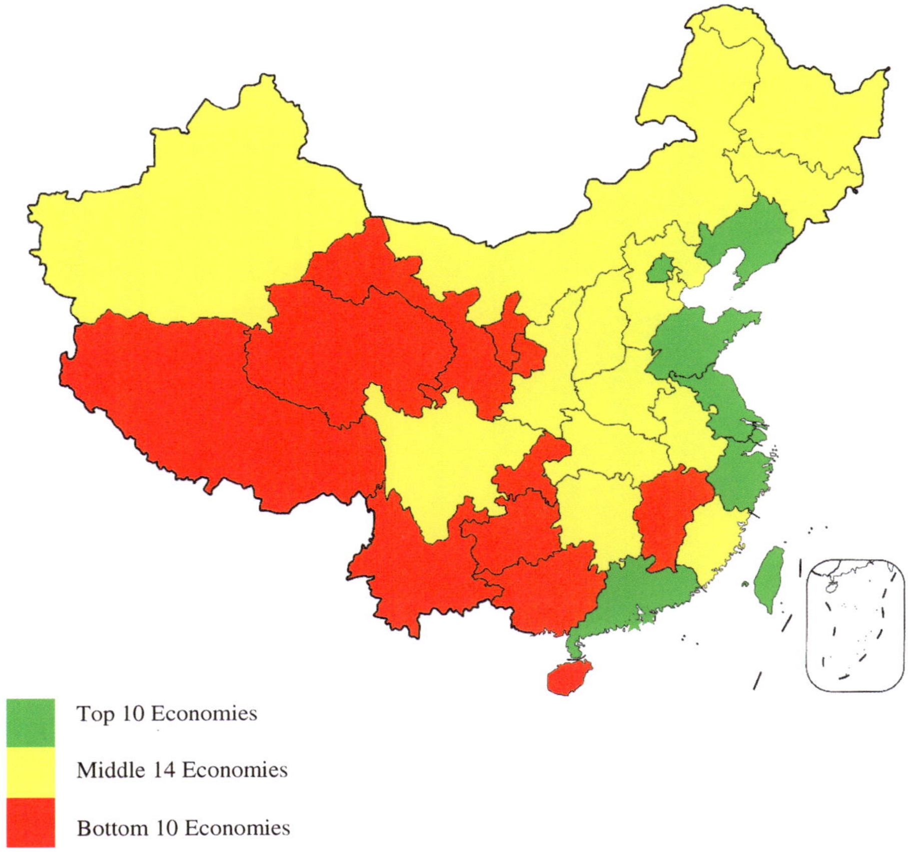

Source: Asia Competitiveness Institute edited from a map provided at http://bbs.eduol.cn/file.php?FileName=20128151347193336.jpg

competitiveness of each region different from each other, economies within the same region have varying competitiveness. For example, Anhui, Henan, Hubei, Hunan and Shanxi from the Central China are in yellow, while Jiangxi from the same region is in red. Significant parts of Western China including Xinjiang, Sichuan, Shaanxi and Inner Mongolia are in yellow instead of red, indicating substantial heterogeneity in competitiveness levels among the economies in Western China.

Table 2.2 illustrates the Overall Competitiveness ranking and score of the 34 Greater China economies from 2000 to 2012. We can see that the Overall Competitiveness rankings of only two economies, namely Beijing and Tibet remain unchanged in 2000 as compared to 2012.

---

includes Heilongjiang, Jilin and Liaoning. The information is from the following website, http://www.stats.gov.cn/tjzs/cjwtjd/201308/t20130829_74318.html (accessed on October 13, 2014) (NBS of People's Republic of China, n.d.).

[9]Result shown in Figure 2.2 is based on 2012 data, collated and analysed in 2015.

Table 2.2: ACI 2015 Overall Competitiveness Ranking and Score: Evolution over Time[10]

	1		2		3		4	
Rank	**Economy**	**Score**	**Economy**	**Score**	**Economy**	**Score**	**Economy**	**Score**
Overall Competitiveness								
2000	Taiwan	3.7611	Hong Kong	2.5729	Guangdong	1.3747	Beijing	1.0387
2001	Taiwan	3.4329	Hong Kong	2.6746	Guangdong	1.5621	Beijing	1.1514
2002	Taiwan	3.3487	Hong Kong	2.3308	Guangdong	1.8125	Shanghai	1.1343
2003	Taiwan	3.1059	Hong Kong	2.4466	Guangdong	1.7865	Shanghai	1.2258
2004	Taiwan	3.0779	Hong Kong	2.3971	Guangdong	1.7245	Beijing	1.2215
2005	Taiwan	2.8879	Hong Kong	2.2869	Guangdong	1.7611	Shanghai	1.3252
2006	Taiwan	2.6819	Hong Kong	2.2270	Guangdong	1.8371	Jiangsu	1.4075
2007	Taiwan	2.4311	Hong Kong	2.2630	Guangdong	1.8006	Beijing	1.4885
2008	Taiwan	2.1890	Hong Kong	2.0626	Guangdong	1.8394	Jiangsu	1.6042
2009	Hong Kong	1.9904	Taiwan	1.9163	Guangdong	1.9078	Jiangsu	1.7735
2010	Guangdong	1.9437	Jiangsu	1.8295	Hong Kong	1.8291	Taiwan	1.8024
2011	Guangdong	1.9334	Jiangsu	1.9105	Hong Kong	1.7876	Shanghai	1.7868
2012	Guangdong	2.0367	Jiangsu	1.9901	Taiwan	1.7056	Beijing	1.6291

(*Continued*)

[10]Result shown in Table 2.2 is based on 2000–2012 data, collated and analysed in 2015.

Table 2.2: (*Continued*)

	5		6		7		8	
Rank	**Economy**	**Score**	**Economy**	**Score**	**Economy**	**Score**	**Economy**	**Score**
2000	Shanghai	0.8759	Jiangsu	0.6214	Liaoning	0.4866	Zhejiang	0.4675
2001	Shanghai	0.9726	Jiangsu	0.6901	Zhejiang	0.6812	Shandong	0.5064
2002	Beijing	1.1023	Jiangsu	0.8600	Zhejiang	0.7043	Shandong	0.6203
2003	Beijing	1.1610	Jiangsu	1.0557	Zhejiang	0.8088	Shandong	0.7090
2004	Shanghai	1.2034	Jiangsu	1.1376	Zhejiang	0.8565	Shandong	0.7644
2005	Jiangsu	1.2938	Beijing	1.2563	Zhejiang	0.9273	Shandong	0.8695
2006	Beijing	1.3264	Shanghai	1.2297	Zhejiang	1.0509	Shandong	1.0132
2007	Jiangsu	1.4793	Shanghai	1.4388	Zhejiang	1.0438	Shandong	0.9140
2008	Beijing	1.5811	Shanghai	1.5111	Zhejiang	1.1070	Shandong	0.9992
2009	Beijing	1.5624	Shanghai	1.5171	Zhejiang	1.1823	Shandong	0.9959
2010	Beijing	1.6114	Shanghai	1.5703	Zhejiang	1.2125	Shandong	0.9374
2011	Taiwan	1.6752	Beijing	1.5854	Zhejiang	1.2317	Shandong	0.8981
2012	Hong Kong	1.6120	Shanghai	1.5189	Zhejiang	1.3244	Shandong	0.9132

	9		10		11		12	
Rank	Economy	Score	Economy	Score	Economy	Score	Economy	Score
2000	Shandong	0.4071	Heilongjiang	–0.0085	Macau	–0.0261	Xinjiang	–0.0322
2001	Liaoning	0.4672	Tianjin	0.0481	Fujian	–0.0352	Macau	–0.0829
2002	Liaoning	0.4778	Macau	0.0764	Fujian	0.0579	Tianjin	0.0308
2003	Liaoning	0.3632	Macau	0.1811	Tianjin	0.0361	Hebei	0.0015
2004	Macau	0.4491	Liaoning	0.2733	Hebei	0.0905	Tianjin	0.0539
2005	Macau	0.3304	Liaoning	0.3024	Tianjin	0.1269	Hebei	0.1107
2006	Macau	0.3347	Liaoning	0.2985	Tianjin	0.1449	Hebei	0.0495
2007	Macau	0.4349	Liaoning	0.2673	Hebei	0.1289	Tianjin	0.0885
2008	Macau	0.3518	Liaoning	0.3425	Tianjin	0.1892	Hebei	0.1486
2009	Liaoning	0.4317	Tianjin	0.2274	Macau	0.2022	Hebei	0.1603
2010	Liaoning	0.4273	Macau	0.4237	Tianjin	0.2290	Hebei	0.1424
2011	Macau	0.4453	Liaoning	0.3864	Fujian	0.1245	Hebei	0.0975
2012	Liaoning	0.4251	Macau	0.4167	Fujian	0.1578	Tianjin	0.0682

(*Continued*)

Table 2.2: *(Continued)*

	13		14		15		16	
Rank	**Economy**	**Score**	**Economy**	**Score**	**Economy**	**Score**	**Economy**	**Score**
2000	Tianjin	−0.0414	Fujian	−0.0822	Hebei	−0.1069	Henan	−0.1417
2001	Hebei	−0.1022	Heilongjiang	−0.1033	Henan	−0.1875	Hubei	−0.2369
2002	Hebei	−0.0368	Heilongjiang	−0.0821	Henan	−0.2260	Jilin	−0.2454
2003	Fujian	−0.0160	Heilongjiang	−0.0909	Xinjiang	−0.2046	Henan	−0.2218
2004	Fujian	−0.0498	Xinjiang	−0.2474	Heilongjiang	−0.2484	Sichuan	−0.2629
2005	Fujian	−0.0369	Heilongjiang	−0.2031	Henan	−0.2348	Xinjiang	−0.2840
2006	Fujian	−0.0305	Henan	−0.1334	Heilongjiang	−0.2501	Inner Mongolia	−0.2592
2007	Fujian	−0.0910	Henan	−0.1306	Xinjiang	−0.2575	Inner Mongolia	−0.2981
2008	Fujian	−0.0157	Henan	−0.1479	Inner Mongolia	−0.2214	Hubei	−0.2867
2009	Fujian	0.0680	Inner Mongolia	−0.1575	Henan	−0.1670	Hubei	−0.2280
2010	Fujian	0.1234	Inner Mongolia	−0.1645	Hubei	−0.1982	Sichuan	−0.2515
2011	Tianjin	0.0955	Inner Mongolia	−0.1427	Hubei	−0.2310	Sichuan	−0.2352
2012	Hebei	0.0595	Sichuan	−0.1626	Hubei	−0.1833	Inner Mongolia	−0.2043

	17		18		19		20	
Rank	**Economy**	**Score**	**Economy**	**Score**	**Economy**	**Score**	**Economy**	**Score**
2000	Hubei	−0.2161	Jilin	−0.2170	Sichuan	−0.2643	Shanxi	−0.2909
2001	Jilin	−0.2430	Xinjiang	−0.2486	Sichuan	−0.2716	Shanxi	−0.3268
2002	Sichuan	−0.2560	Xinjiang	−0.2781	Shanxi	−0.3566	Hubei	−0.3732
2003	Sichuan	−0.2360	Hubei	−0.2794	Jilin	−0.3477	Shanxi	−0.4007
2004	Henan	−0.2738	Hubei	−0.3219	Inner Mongolia	−0.4092	Jilin	−0.4288
2005	Inner Mongolia	−0.3047	Hubei	−0.3589	Sichuan	−0.3926	Jilin	−0.4425
2006	Xinjiang	−0.2887	Hubei	−0.3244	Sichuan	−0.3281	Jilin	−0.4022
2007	Heilongjiang	−0.3119	Sichuan	−0.3250	Hubei	−0.3369	Jilin	−0.4126
2008	Heilongjiang	−0.2983	Xinjiang	−0.3024	Sichuan	−0.4016	Jilin	−0.4122
2009	Sichuan	−0.3064	Jilin	−0.3513	Heilongjiang	−0.3571	Xinjiang	−0.4109
2010	Henan	−0.2630	Xinjiang	−0.3042	Heilongjiang	−0.3470	Shaanxi	−0.4510
2011	Xinjiang	−0.3115	Heilongjiang	−0.3224	Henan	−0.3264	Anhui	−0.4601
2012	Henan	−0.2842	Shaanxi	−0.3299	Anhui	−0.3646	Xinjiang	−0.3816

(Continued)

Table 2.2: *(Continued)*

	21		22		23		24	
Rank	**Economy**	**Score**	**Economy**	**Score**	**Economy**	**Score**	**Economy**	**Score**
2000	Inner Mongolia	–0.3241	Yunnan	–0.3734	Ningxia	–0.5285	Jiangxi	–0.6177
2001	Yunnan	–0.3681	Inner Mongolia	–0.3745	Ningxia	–0.5525	Jiangxi	–0.5981
2002	Hunan	–0.3873	Yunnan	–0.4579	Inner Mongolia	–0.4914	Chongqing	–0.5619
2003	Inner Mongolia	–0.4460	Yunnan	–0.5745	Ningxia	–0.6171	Jiangxi	–0.6723
2004	Shanxi	–0.4559	Hainan	–0.6250	Yunnan	–0.6431	Ningxia	–0.6611
2005	Shanxi	–0.4545	Yunnan	–0.6037	Hainan	–0.6431	Shaanxi	–0.6758
2006	Shanxi	–0.5905	Hunan	–0.6043	Shaanxi	–0.6096	Yunnan	–0.6638
2007	Shanxi	–0.4853	Hunan	–0.5769	Shaanxi	–0.6445	Jiangxi	–0.6836
2008	Shanxi	–0.4666	Shaanxi	–0.5625	Hunan	–0.5944	Jiangxi	–0.6645
2009	Shaanxi	–0.5084	Shanxi	–0.5249	Anhui	–0.5459	Hunan	–0.5770
2010	Anhui	–0.5070	Jilin	–0.5157	Hunan	–0.5338	Shanxi	–0.5397
2011	Jilin	–0.4684	Shaanxi	–0.4911	Shanxi	–0.5453	Hunan	–0.5875
2012	Jilin	–0.4076	Heilongjiang	–0.4317	Hunan	–0.5517	Shanxi	–0.5588

Rank	25		26		27		28	
	Economy	Score	Economy	Score	Economy	Score	Economy	Score
2000	Qinghai	–0.6613	Hunan	–0.6771	Guangxi	–0.7048	Chongqing	–0.7048
2001	Hunan	–0.6377	Guangxi	–0.6899	Chongqing	–0.7209	Qinghai	–0.7445
2002	Ningxia	–0.5922	Jiangxi	–0.6633	Qinghai	–0.7355	Anhui	–0.8020
2003	Chongqing	–0.6821	Hunan	–0.6954	Anhui	–0.7501	Shaanxi	–0.7520
2004	Jiangxi	–0.6765	Hunan	–0.6765	Shaanxi	–0.7101	Qinghai	–0.7557
2005	Hunan	–0.6772	Jiangxi	–0.7011	Ningxia	–0.7261	Chongqing	–0.7411
2006	Jiangxi	–0.6830	Anhui	–0.7572	Chongqing	–0.7850	Guangxi	–0.8028
2007	Yunnan	–0.6945	Anhui	–0.7255	Chongqing	–0.7793	Guangxi	–0.7941
2008	Yunnan	–0.6699	Anhui	–0.6981	Chongqing	–0.7875	Guangxi	–0.7924
2009	Jiangxi	–0.6443	Chongqing	–0.7345	Yunnan	–0.7349	Guangxi	–0.7713
2010	Jiangxi	–0.7249	Yunnan	–0.7336	Chongqing	–0.7447	Guangxi	–0.8078
2011	Chongqing	–0.6679	Yunnan	–0.6935	Jiangxi	–0.7165	Guangxi	–0.7631
2012	Yunnan	–0.6348	Chongqing	–0.6536	Jiangxi	–0.7870	Hainan	–0.8450

(Continued)

Table 2.2: (*Continued*)

Rank	29		30		31		32	
	Economy	**Score**	**Economy**	**Score**	**Economy**	**Score**	**Economy**	**Score**
2000	Shaanxi	−0.7130	Hainan	−0.8254	Anhui	−0.8492	Gansu	−0.8674
2001	Shaanxi	−0.7525	Anhui	−0.8031	Gansu	−0.8214	Hainan	−0.8402
2002	Shaanxi	−0.8047	Guangxi	−0.8300	Hainan	−0.8723	Gansu	−0.9810
2003	Qinghai	−0.7660	Guangxi	−0.8816	Hainan	−0.8990	Gansu	−0.9897
2004	Anhui	−0.7627	Chongqing	−0.8122	Guangxi	−0.8432	Gansu	−0.9906
2005	Qinghai	−0.7495	Anhui	−0.8267	Guangxi	−0.8750	Gansu	−1.0465
2006	Ningxia	−0.8236	Qinghai	−0.8359	Hainan	−0.8808	Gansu	−1.0006
2007	Ningxia	−0.8455	Hainan	−0.8713	Qinghai	−0.9039	Gansu	−1.0920
2008	Ningxia	−0.8599	Hainan	−0.9296	Qinghai	−0.9635	Guizhou	−1.1762
2009	Hainan	−0.9033	Ningxia	−0.9229	Qinghai	−1.0684	Guizhou	−1.1772
2010	Hainan	−0.8705	Ningxia	−0.9359	Qinghai	−1.1431	Guizhou	−1.2337
2011	Hainan	−0.8689	Ningxia	−0.9545	Qinghai	−1.1288	Gansu	−1.2307
2012	Guangxi	−0.8529	Ningxia	−0.9870	Qinghai	−1.1515	Guizhou	−1.1619

	33		34	
Rank	**Economy**	**Score**	**Economy**	**Score**
2000	Guizhou	−1.0569	Tibet	−1.2753
2001	Guizhou	−1.0914	Tibet	−1.3537
2002	Guizhou	−1.1078	Tibet	−1.4145
2003	Guizhou	−1.1514	Tibet	−1.2073
2004	Guizhou	−1.1659	Tibet	−1.2289
2005	Guizhou	−1.1386	Tibet	−1.3619
2006	Guizhou	−1.1475	Tibet	−1.4003
2007	Guizhou	−1.1573	Tibet	−1.3618
2008	Gansu	−1.2311	Tibet	−1.4431
2009	Gansu	−1.3035	Tibet	−1.5403
2010	Gansu	−1.3153	Tibet	−1.4972
2011	Guizhou	−1.3236	Tibet	−1.4893
2012	Gansu	−1.2443	Tibet	−1.6789

Source: Asia Competitiveness Institute

Meanwhile, the rankings of 17 economies have improved in the same period (2000–2012). Guizhou, Macau, Shandong, Tianjin and Zhejiang each rises one spot; Chongqing, Guangdong, Hainan, Hebei and Hubei each rises two spots; Fujian and Hunan each rises three spots; Jiangsu rises four spots; and Sichuan and Inner Mongolia rises five spots. There are also 15 economies whose rankings fall from 2000 to 2012. Gansu, Henan and Shanghai each falls one spot; Guangxi, Liaoning and Taiwan each falls two spots; Hong Kong, Jiangxi, Jilin and Yunnan each falls three spots; Shanxi falls four spots; Qinghai falls six spots; Ningxia falls seven spots; and Xinjiang falls eight spots.

Three mainland China economies have experienced substantial changes in their Overall Competitiveness ranking over the past 13-year period from 2000 to 2012. In particular, the ranking of Heilongjiang has dropped by 12 places while the rankings of Anhui and Shaanxi have improved by 12 and 11 places, respectively. Interestingly, these three economies are located in three different regions in the mainland China, namely Anhui from Central China, Heilongjiang from North-eastern China, and Shaanxi from Western China, indicating that some major changes have occurred across the competitiveness landscape of mainland China over the past decade or so. Such changes are by no means random fluctuations but can be linked to mainland China's regional development strategies, which were discussed in Chapters 4–7 in "2014 Annual Provincial and Regional Competitiveness Analysis for Greater China Economies: Development Strategies under the New Normal", the previous volume of ACI's annual analysis of competitiveness on Greater China (Tan, Yuan and Yoong 2015).

The gap in competitiveness between economies in mainland China and Hong Kong and Taiwan has diminished from 2000 to 2009. In 2000, Taiwan and Hong Kong scored 3.7611 and 2.5729, respectively for Overall Competitiveness, while Guangdong from the mainland China achieved a score of 1.3747. However, in 2009, the scores of Hong Kong and Taiwan have decreased substantially to 1.9904 and 1.9163, respectively while the score of Guangdong has improved to 1.9078. This means that the competition to remain top within Greater China has intensified over the period from 2000 to 2009. Since 2010, economies from the mainland China like Guangdong and Jiangsu have surpassed Taiwan and Hong Kong in terms of Overall Competitiveness.

2.2.2 ACI 2015 Macroeconomic Stability Ranking and Score

Table 2.3 shows the 2015 Macroeconomic Stability ranking. Jiangsu has come first in the Macroeconomic Stability ranking with a score of 2.6472. On the other hand, Macau is ranked the lowest among the 34 Greater China economies with a score of –1.2874.

The average performing economy (with a standardised score of 0), for Macroeconomic Stability ranking, is somewhere between the #12-ranked Tianjin (0.0286) and the #13-ranked Hubei (–0.0652). The median which is at the middle of a ranking, it is situated between #17-ranked Hunan (–0.2612) and the #18-ranked Chongqing (–0.2820). For Macroeconomic Stability, the median score is –0.2716. Similar to the case of Overall Competitiveness ranking, the average score (0) of Macroeconomic Stability ranking is higher than the median score (–0.2716).

Table 2.3: ACI 2015 Macroeconomic Stability Ranking and Score[11]

Rank	Economy	Score
1	Jiangsu	2.6472
2	Guangdong	2.6039
3	Shanghai	1.7310
4	Hong Kong	1.5224
5	Shandong	1.2727
6	Zhejiang	1.0992
7	Liaoning	0.9756
8	Beijing	0.6305
9	Taiwan	0.3778
10	Fujian	0.2505
11	Sichuan	0.0813
12	Tianjin	0.0286
13	Hubei	–0.0652
14	Henan	–0.1272
15	Hebei	–0.1345
16	Anhui	–0.1488
17	Hunan	–0.2612
18	Chongqing	–0.2820
19	Shaanxi	–0.3474
20	Jilin	–0.4919
21	Yunnan	–0.4970
22	Guangxi	–0.5169
23	Jiangxi	–0.5920
24	Heilongjiang	–0.6104
25	Inner Mongolia	–0.6436
26	Guizhou	–0.7109
27	Shanxi	–0.7260
28	Tibet	–0.8492
29	Xinjiang	–0.8958
30	Hainan	–0.9530
31	Ningxia	−1.0166
32	Qinghai	−1.0246
33	Gansu	−1.0392
34	Macau	−1.2874

Source: Asia Competitiveness Institute

[11] Result shown in Table 2.3 is based on 2012 data, collated and analysed in 2015.

Figure 2.3: ACI 2015 Macroeconomic Stability Ranking Map[12]

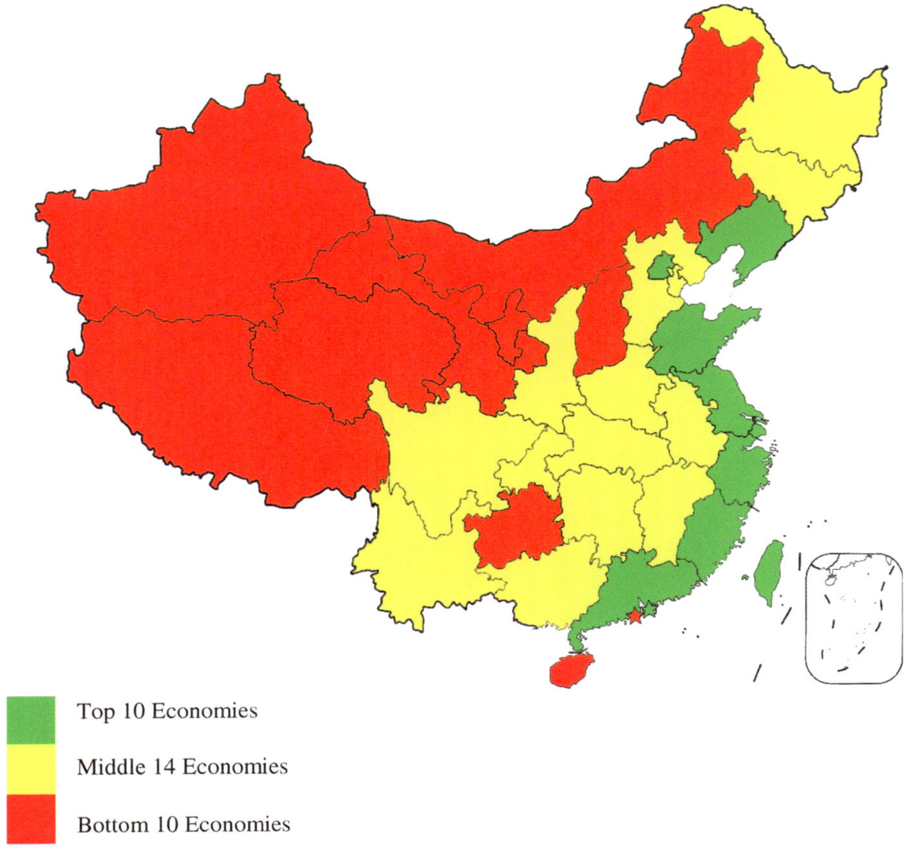

Source: Asia Competitiveness Institute edited from a map provided at http://bbs.eduol.cn/file.php?FileName=20128151347193336.jpg

It is worth noting that there is a marked drop in score from the #2-ranked Guangdong (2.6039) to the #3-ranked Shanghai (1.7310). The gap in scores between these two economies is 0.8729. This means that the top two ranked economies have considerably outperformed the rest of the 34 Greater China economies in terms of Macroeconomic Stability. Another larger score gap is observed between #7-ranked Liaoning (0.9756) and #8-ranked Beijing (0.6305), whereby their score difference is 0.3451.

Figure 2.3 provides a snapshot for the 2015 Macroeconomic Stability ranking. Similar to Figure 2.2, green represents the top 10 economies, yellow, the middle 14 economies and red, the 10 lowest ranking economies. We observe that the coastal area in mainland China is largely green, except for Hebei and Tianjin. In general, the colour of the economy changes from green, to yellow, then to red as we move westwards towards inland areas in mainland China. Some economies that differ noticeably from this general pattern observed are Jilin, Shanxi, Guizhou, Hainan and Macau.

Table 2.4 shows the Macroeconomic Stability ranking and scores for the 34 Greater China economies from 2000 to 2012. Among the 34 economies, the rankings of Guangdong, Guizhou,

[12]Result shown in Figure 2.3 is based on 2012 data, collated and analysed in 2015.

Table 2.4: ACI 2015 Macroeconomic Stability Ranking and Score: Evolution over Time[13]

	1		2		3		4	
Rank	**Economy**	**Score**	**Economy**	**Score**	**Economy**	**Score**	**Economy**	**Score**
Macroeconomic Stability								
2000	Hong Kong	3.5878	Guangdong	1.9863	Shanghai	1.7522	Taiwan	1.5178
2001	Hong Kong	3.2233	Guangdong	2.4564	Shanghai	1.8032	Jiangsu	1.3785
2002	Guangdong	2.7825	Hong Kong	2.2895	Shanghai	2.0755	Jiangsu	1.7081
2003	Guangdong	2.6911	Hong Kong	2.3797	Shanghai	2.1332	Jiangsu	2.0003
2004	Guangdong	2.5699	Hong Kong	2.4926	Shanghai	2.1383	Jiangsu	1.9690
2005	Guangdong	2.4368	Hong Kong	2.4260	Jiangsu	2.1691	Shanghai	2.0601
2006	Guangdong	2.4713	Jiangsu	2.2329	Hong Kong	2.1420	Shanghai	1.9391
2007	Guangdong	2.3921	Hong Kong	2.3286	Jiangsu	2.2620	Shanghai	1.9016
2008	Jiangsu	2.4400	Guangdong	2.3840	Hong Kong	1.9755	Shanghai	1.9405
2009	Jiangsu	2.5451	Guangdong	2.4859	Hong Kong	1.8403	Shanghai	1.8395
2010	Jiangsu	2.4905	Guangdong	2.4548	Shanghai	1.8672	Hong Kong	1.7909
2011	Jiangsu	2.6210	Guangdong	2.4687	Shanghai	1.8023	Hong Kong	1.7604
2012	Jiangsu	2.6472	Guangdong	2.6039	Shanghai	1.7310	Hong Kong	1.5224

(Continued)

[13] Result shown in Table 2.4 is based on 2000–2012 data, collated and analysed in 2015.

Table 2.4: *(Continued)*

	5		6		7		8	
Rank	**Economy**	**Score**	**Economy**	**Score**	**Economy**	**Score**	**Economy**	**Score**
2000	Jiangsu	1.3539	Beijing	1.1215	Shandong	0.4173	Zhejiang	0.3321
2001	Taiwan	1.2675	Beijing	1.1145	Zhejiang	0.6069	Shandong	0.5655
2002	Taiwan	1.2029	Beijing	1.0208	Zhejiang	0.8384	Shandong	0.8346
2003	Shandong	1.0150	Taiwan	0.9943	Zhejiang	0.9217	Beijing	0.7893
2004	Shandong	1.0438	Taiwan	1.0406	Beijing	0.9014	Zhejiang	0.8826
2005	Shandong	1.1369	Zhejiang	0.9635	Taiwan	0.7977	Beijing	0.7605
2006	Shandong	1.2627	Zhejiang	1.1319	Taiwan	0.8900	Beijing	0.8651
2007	Shandong	1.1483	Zhejiang	1.0705	Beijing	0.9955	Taiwan	0.7550
2008	Shandong	1.2572	Zhejiang	1.0847	Beijing	0.8832	Liaoning	0.7362
2009	Shandong	1.2156	Zhejiang	1.1332	Liaoning	0.8448	Beijing	0.6152
2010	Shandong	1.2996	Zhejiang	1.0566	Liaoning	0.9772	Beijing	0.7469
2011	Shandong	1.2657	Zhejiang	1.0652	Liaoning	0.9226	Beijing	0.6561
2012	Shandong	1.2727	Zhejiang	1.0992	Liaoning	0.9756	Beijing	0.6305

	9		10		11		12	
Rank	**Economy**	**Score**	**Economy**	**Score**	**Economy**	**Score**	**Economy**	**Score**
2000	Liaoning	0.2760	Tianjin	0.1159	Fujian	0.1060	Hubei	−0.1345
2001	Fujian	0.3089	Liaoning	0.2743	Tianjin	0.1535	Hebei	−0.1632
2002	Fujian	0.4276	Liaoning	0.3918	Tianjin	0.1906	Hebei	−0.1676
2003	Liaoning	0.2696	Fujian	0.2295	Tianjin	0.1619	Hebei	−0.1980
2004	Tianjin	0.2427	Liaoning	0.1815	Fujian	0.1131	Hebei	−0.1282
2005	Liaoning	0.4599	Tianjin	0.3612	Fujian	0.0859	Hebei	−0.1324
2006	Liaoning	0.4707	Tianjin	0.2266	Fujian	0.1724	Henan	−0.1402
2007	Liaoning	0.5134	Tianjin	0.1801	Fujian	0.1276	Hebei	−0.1303
2008	Taiwan	0.5115	Tianjin	0.2484	Fujian	0.1560	Hebei	−0.1136
2009	Fujian	0.2864	Taiwan	0.2779	Tianjin	0.1494	Sichuan	−0.0389
2010	Taiwan	0.4980	Fujian	0.2035	Tianjin	0.0630	Sichuan	0.0000
2011	Taiwan	0.4715	Fujian	0.1994	Tianjin	0.0593	Sichuan	0.0198
2012	Taiwan	0.3778	Fujian	0.2505	Sichuan	0.0813	Tianjin	0.0286

(Continued)

Table 2.4: *(Continued)*

	13		14		15		16	
Rank	**Economy**	**Score**	**Economy**	**Score**	**Economy**	**Score**	**Economy**	**Score**
2000	Henan	–0.1438	Hebei	–0.1452	Guangxi	–0.2176	Chongqing	–0.2998
2001	Hubei	–0.1765	Guangxi	–0.1883	Sichuan	–0.3951	Jiangxi	–0.4258
2002	Hubei	–0.1901	Guangxi	–0.2896	Sichuan	–0.3156	Anhui	–0.4311
2003	Sichuan	–0.2915	Henan	–0.3119	Hubei	–0.3379	Guangxi	–0.4067
2004	Sichuan	–0.2971	Henan	–0.3325	Hubei	–0.3916	Guangxi	–0.4252
2005	Henan	–0.2064	Sichuan	–0.2498	Chongqing	–0.3171	Hubei	–0.4298
2006	Hebei	–0.1578	Sichuan	–0.2073	Hubei	–0.3472	Guangxi	–0.4049
2007	Henan	–0.1374	Sichuan	–0.1729	Hubei	–0.2654	Anhui	–0.4239
2008	Sichuan	–0.1181	Henan	–0.1560	Hubei	–0.2722	Anhui	–0.3569
2009	Anhui	–0.0590	Hebei	–0.0606	Hubei	–0.1620	Henan	–0.1701
2010	Hebei	–0.0904	Hubei	–0.0957	Henan	–0.1731	Anhui	–0.2515
2011	Hubei	–0.0753	Hebei	–0.0840	Henan	–0.1646	Anhui	–0.2622
2012	Hubei	–0.0652	Henan	–0.1272	Hebei	–0.1345	Anhui	–0.1488

Rank	17		18		19		20	
	Economy	Score	Economy	Score	Economy	Score	Economy	Score
2000	Sichuan	–0.3476	Shaanxi	–0.3846	Heilongjiang	–0.4027	Jilin	–0.4419
2001	Shaanxi	–0.4341	Anhui	–0.4661	Yunnan	–0.4675	Henan	–0.4786
2002	Shanxi	–0.4478	Shaanxi	–0.4833	Henan	–0.4973	Hunan	–0.5098
2003	Jiangxi	–0.4671	Chongqing	–0.4859	Anhui	–0.5202	Jilin	–0.5495
2004	Jiangxi	–0.4413	Anhui	–0.4669	Chongqing	–0.5154	Shaanxi	–0.5448
2005	Shaanxi	–0.4407	Anhui	–0.4839	Guangxi	–0.4925	Jiangxi	–0.5161
2006	Anhui	–0.4319	Jiangxi	–0.4632	Inner Mongolia	–0.4634	Hunan	–0.4684
2007	Hunan	–0.4259	Jilin	–0.4392	Inner Mongolia	–0.4407	Guangxi	–0.4786
2008	Hunan	–0.3762	Jilin	–0.3926	Inner Mongolia	–0.4121	Chongqing	–0.4313
2009	Chongqing	–0.3311	Guangxi	–0.3508	Hunan	–0.3517	Jilin	–0.4257
2010	Hunan	–0.3192	Chongqing	–0.3719	Guangxi	–0.3731	Jiangxi	–0.4662
2011	Chongqing	–0.2965	Hunan	–0.3047	Guangxi	–0.4536	Shaanxi	–0.4697
2012	Hunan	–0.2612	Chongqing	–0.2820	Shaanxi	–0.3474	Jilin	–0.4919

(*Continued*)

Table 2.4: (*Continued*)

	21		22		23		24	
Rank	**Economy**	**Score**	**Economy**	**Score**	**Economy**	**Score**	**Economy**	**Score**
2000	Yunnan	−0.5011	Anhui	−0.5178	Jiangxi	−0.5667	Xinjiang	−0.5686
2001	Hunan	−0.4924	Chongqing	−0.5452	Jilin	−0.5493	Hainan	−0.5519
2002	Jiangxi	−0.5202	Chongqing	−0.5305	Yunnan	−0.5523	Jilin	−0.5695
2003	Hunan	−0.5755	Heilongjiang	−0.5764	Shanxi	−0.5796	Yunnan	−0.5805
2004	Hunan	−0.5460	Inner Mongolia	−0.5600	Heilongjiang	−0.6186	Shanxi	−0.6458
2005	Inner Mongolia	−0.5315	Heilongjiang	−0.5509	Hunan	−0.5712	Jilin	−0.5736
2006	Jilin	−0.4761	Shaanxi	−0.5149	Chongqing	−0.6324	Heilongjiang	−0.6443
2007	Jiangxi	−0.5263	Chongqing	−0.5434	Shaanxi	−0.5493	Shanxi	−0.6319
2008	Jiangxi	−0.4770	Guangxi	−0.4989	Shaanxi	−0.5003	Heilongjiang	−0.5092
2009	Jiangxi	−0.4358	Inner Mongolia	−0.4592	Shaanxi	−0.5413	Yunnan	−0.6079
2010	Shaanxi	−0.5275	Inner Mongolia	−0.5406	Heilongjiang	−0.5713	Jilin	−0.5757
2011	Jiangxi	−0.4986	Yunnan	−0.5075	Inner Mongolia	−0.5268	Heilongjiang	−0.5490
2012	Yunnan	−0.4970	Guangxi	−0.5169	Jiangxi	−0.5920	Heilongjiang	−0.6104

	25		26		27		28	
Rank	**Economy**	**Score**	**Economy**	**Score**	**Economy**	**Score**	**Economy**	**Score**
2000	Hunan	–0.6380	Guizhou	–0.6602	Ningxia	–0.6888	Hainan	–0.6958
2001	Shanxi	–0.5713	Heilongjiang	–0.6102	Guizhou	–0.6746	Ningxia	–0.7449
2002	Heilongjiang	–0.6100	Hainan	–0.6212	Guizhou	–0.7616	Ningxia	–0.8180
2003	Shaanxi	–0.5940	Xinjiang	–0.6718	Inner Mongolia	–0.6755	Hainan	–0.7413
2004	Jilin	–0.6473	Yunnan	–0.6712	Macau	–0.6812	Xinjiang	–0.7219
2005	Yunnan	–0.6378	Xinjiang	–0.6600	Shanxi	–0.7165	Guizhou	–0.7544
2006	Yunnan	–0.6502	Shanxi	–0.7455	Xinjiang	–0.7581	Hainan	–0.7958
2007	Heilongjiang	–0.6554	Yunnan	–0.6669	Tibet	–0.7781	Hainan	–0.8195
2008	Yunnan	–0.6166	Shanxi	–0.7017	Xinjiang	–0.8355	Tibet	–0.8375
2009	Heilongjiang	–0.6102	Shanxi	–0.7443	Guizhou	–0.7597	Hainan	–0.8301
2010	Shanxi	–0.6176	Yunnan	–0.6687	Guizhou	–0.8597	Xinjiang	–0.8719
2011	Jilin	–0.5795	Shanxi	–0.6587	Guizhou	–0.8247	Xinjiang	–0.8857
2012	Inner Mongolia	–0.6436	Guizhou	–0.7109	Shanxi	–0.7260	Tibet	–0.8492

(*Continued*)

Table 2.4: (*Continued*)

Rank	29		30		31		32	
	Economy	Score	Economy	Score	Economy	Score	Economy	Score
2000	Shanxi	−0.7952	Qinghai	−0.8206	Inner Mongolia	−0.8540	Gansu	−0.8963
2001	Inner Mongolia	−0.7812	Qinghai	−0.8267	Tibet	−0.8685	Gansu	−0.8811
2002	Xinjiang	−0.8281	Inner Mongolia	−0.8368	Macau	−0.8928	Gansu	−0.9503
2003	Guizhou	−0.7471	Ningxia	−0.7619	Macau	−0.8552	Gansu	−0.8623
2004	Tibet	−0.7811	Gansu	−0.8002	Ningxia	−0.8153	Guizhou	−0.8193
2005	Qinghai	−0.8283	Hainan	−0.8320	Ningxia	−0.8335	Tibet	−0.8354
2006	Guizhou	−0.8195	Ningxia	−0.8470	Tibet	−0.8571	Qinghai	−0.8952
2007	Xinjiang	−0.8262	Ningxia	−0.8619	Guizhou	−0.8715	Gansu	−0.9892
2008	Guizhou	−0.8580	Hainan	−0.8590	Ningxia	−0.9523	Gansu	−1.0875
2009	Tibet	−0.9249	Xinjiang	−0.9318	Ningxia	−0.9646	Gansu	−1.1101
2010	Tibet	−0.8767	Macau	−0.9532	Hainan	−0.9719	Ningxia	−1.0138
2011	Tibet	−0.9029	Hainan	−0.9941	Macau	−1.0602	Ningxia	−1.0673
2012	Xinjiang	−0.8958	Hainan	−0.9530	Ningxia	−1.0166	Qinghai	−1.0246

Rank	33		34	
	Economy	**Score**	**Economy**	**Score**
2000	Macau	–0.9070	Tibet	–0.9392
2001	Macau	–0.9144	Xinjiang	–0.9454
2002	Tibet	–0.9570	Qinghai	–0.9820
2003	Tibet	–0.8838	Qinghai	–0.9118
2004	Qinghai	–0.8462	Hainan	–0.8783
2005	Gansu	–0.9066	Macau	–1.1570
2006	Gansu	–0.8994	Macau	–1.1853
2007	Macau	–1.0042	Qinghai	–1.0367
2008	Macau	–1.1197	Qinghai	–1.1350
2009	Macau	–1.1792	Qinghai	–1.1842
2010	Gansu	–1.0979	Qinghai	–1.1607
2011	Gansu	–1.0711	Qinghai	–1.0749
2012	Gansu	–1.0392	Macau	–1.2874

Source: Asia Competitiveness Institute

Jiangxi, Jilin, Shanghai and Yunnan did not change in 2000 as compared to 2012. Meanwhile, the rankings of nine economies have improved over the years. Fujian increases one place; Liaoning and Shanxi each increases two places; Jiangsu increases four places; Anhui, Inner Mongolia, Sichuan and Tibet each increases six places; and Hunan increases eight places. The rankings of the remaining 19 economies have decreased over the years. Gansu, Hebei, Henan, Hubei, Macau and Shaanxi each decreases one place; Beijing, Chongqing, Hainan, Qinghai, Shandong, Tianjin and Zhejiang each decreases two places; Hong Kong decreases three places; Ningxia decreases four places; Heilongjiang, Taiwan and Xinjiang each decreases five places; and Guangxi decreases seven places.

Hong Kong only retained its number one position in the Macroeconomic Stability ranking for 2000 and 2001, and its standardised score has been decreasing steadily over the past 13 years, from 3.5878 in 2000 to 1.5224 in 2012. Beginning 2003, Hong Kong relinquished its top position in Macroeconomic Stability to mainland China economies such Guangdong, Jiangsu and Shanghai, and has not regained the top position since then. The gap between the top ranked economy and the economy in the second position is getting smaller, whereby the difference in scores has reduced from 1.6015 between Hong Kong and Guangdong in 2000 to a mere 0.0433 between Jiangsu and Guangdong in 2012. The competition to be at the top of the Macroeconomic Stability ranking has certainly been stronger, indicating the intensifying rivalry and dynamism among economies from the mainland in areas related to this environment.

2.2.3 ACI 2015 Government and Institutional Setting Ranking and Score

The Government and Institutional Setting ranking is displayed in Table 2.5. The maximum and minimum score of the 34 Greater China economies are 2.2121 and –2.0110, respectively. Taiwan occupies the highest position in this environment while Tibet is ranked at the bottom.

The average-performing economy is situated between the #9-ranked Shandong (0.7651) and the #10-ranked Fujian (–0.0487), while the median score lies between the #17-ranked Yunnan (–0.3252) and the #18-ranked Anhui (–0.3281). For the Government and Institutional Setting, the median score is –0.3267. The median score (–0.3267) is noticeably lower than the average score (0).

The distribution of scores for Government and Institutional Setting is quite skewed, with only nine economies out of the 34 Greater China economies having scores above the average (i.e. 0). The difference in scores for the bottom ranked Tibet (–2.0110) and the #33-ranked Gansu (–1.0033) is 1.0077. Tibet can be viewed as an outlier in the Government and Institutional Setting environment. Also, the gap in scores between the #9-ranked Shandong (0.7651) and the #10-ranked Fujian (–0.0487) is 0.8138.

Figure 2.4 shows the 2015 Government and Institutional Setting ranking from a spatial perspective. The top 10 economies are colour-coded in green, middle 14 economies in yellow, and bottom 10 in red. We observe that the coastal area in mainland China is largely green, except for Hebei and Tianjin. The distribution of the economies in yellow and red lacks clearly defined patterns, as these middle or bottom-ranked economies for this environment span across all the regions of mainland China with the exception of the Eastern Coastal Area where most

Table 2.5: ACI 2015 Government and Institutional Setting Ranking and Score[14]

Rank	Economy	Score
1	Taiwan	2.2121
2	Jiangsu	1.8222
3	Guangdong	1.6910
4	Beijing	1.6866
5	Hong Kong	1.6646
6	Shanghai	1.4031
7	Zhejiang	1.3717
8	Macau	1.1074
9	Shandong	0.7651
10	Fujian	–0.0487
11	Liaoning	–0.0705
12	Sichuan	–0.1431
13	Hebei	–0.1845
14	Inner Mongolia	–0.2327
15	Xinjiang	–0.2358
16	Hubei	–0.2516
17	Yunnan	–0.3252
18	Anhui	–0.3281
19	Henan	–0.3777
20	Hainan	–0.3992
21	Heilongjiang	–0.4812
22	Shaanxi	–0.4836
23	Jiangxi	–0.4928
24	Tianjin	–0.5133
25	Ningxia	–0.5503
26	Chongqing	–0.6368
27	Jilin	–0.6383
28	Hunan	–0.7580
29	Guangxi	–0.8003
30	Shanxi	–0.8516
31	Guizhou	–0.9100
32	Qinghai	–0.9960
33	Gansu	–1.0033
34	Tibet	–2.0110

Source: Asia Competitiveness Institute

[14]Result shown in Table 2.5 is based on 2012 data, collated and analysed in 2015.

Figure 2.4: ACI 2015 Government and Institutional Setting Ranking Map[15]

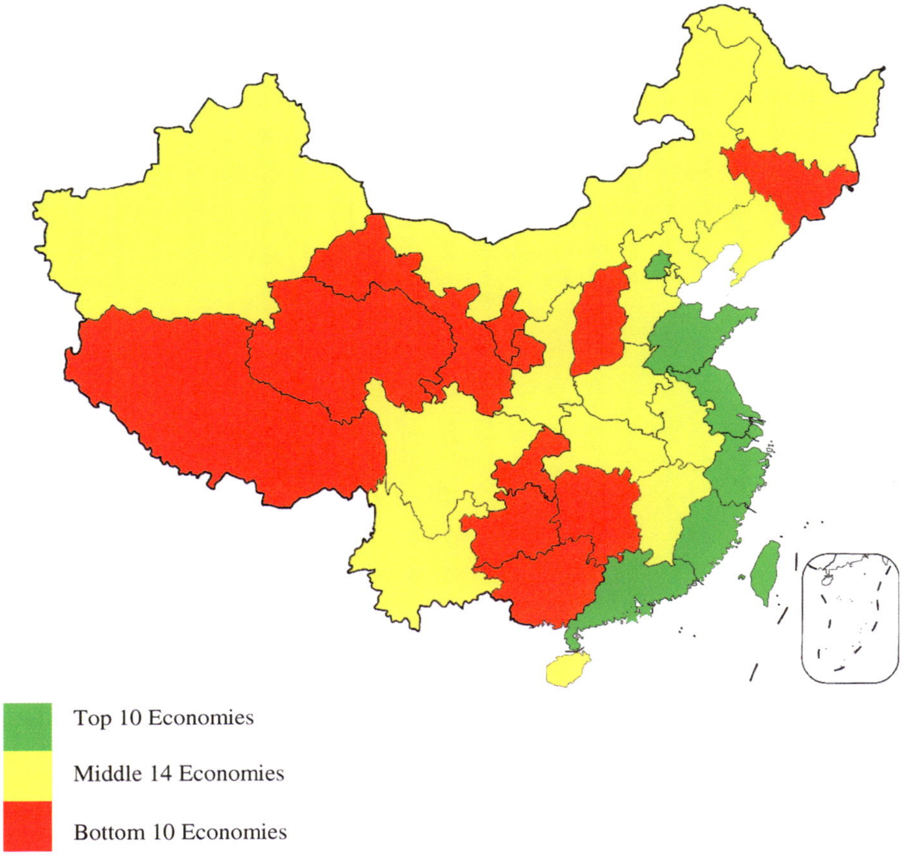

Source: Asia Competitiveness Institute edited from a map provided at http://bbs.eduol.cn/file.php?FileName=20128151347193336.jpg

economies are in the top 10. That said, the bottom 10 economies do tend to concentrate in two clusters that consist of Tibet, Qinghai Gansu and Ningxia in the west and Guizhou, Chongqing, Hunan and Guangxi in the south, respectively.

Table 2.6 shows the Government and Institutional Setting ranking and scores for the 34 Greater China economies from 2000 to 2012. There are three economies whose ranking in 2012 remain the same as in 2000. They are Guangdong, Taiwan and Tibet. There are 15 economies with improved rankings over the past 13 years from 2000 to 2012. Guangxi rises one spot; Jiangsu and Hunan each rises four spots; Beijing, Fujian, Inner Mongolia and Macau each rises five spots; Hainan and Jiangxi each rises six spots; Sichuan rises eight spots; Hubei and Shaanxi each rises 11 spots; Hebei rises 12 spots; and Anhui rises 13 spots. On the other hand, the ranking of 17 economies have fallen over the years. Liaoning and Shanghai each falls one spot; Chongqing and Shandong each falls two spots; Hong Kong, Tianjin and Zhejiang each falls three

[15]Result shown in Figure 2.4 is based on 2012 data, collated and analysed in 2015.

Table 2.6: ACI 2015 Government and Institutional Setting Ranking and Score: Evolution over Time[16]

	1		2		3		4	
Rank	**Economy**	**Score**	**Economy**	**Score**	**Economy**	**Score**	**Economy**	**Score**
Government and Institutional Setting								
2000	Taiwan	4.2982	Hong Kong	1.4489	Guangdong	1.0755	Zhejiang	0.9308
2001	Taiwan	3.7941	Hong Kong	1.8538	Zhejiang	1.2793	Guangdong	1.2710
2002	Taiwan	3.6941	Hong Kong	1.7099	Guangdong	1.3416	Zhejiang	1.1687
2003	Taiwan	3.6474	Hong Kong	1.9338	Guangdong	1.3733	Zhejiang	1.2115
2004	Taiwan	3.5354	Hong Kong	2.0065	Guangdong	1.2588	Beijing	1.2225
2005	Taiwan	3.3843	Hong Kong	2.0840	Beijing	1.3829	Shanghai	1.3176
2006	Taiwan	3.2369	Hong Kong	2.0987	Beijing	1.3653	Guangdong	1.2423
2007	Taiwan	2.9589	Hong Kong	2.1014	Beijing	1.4194	Guangdong	1.3588
2008	Taiwan	2.8100	Hong Kong	1.8650	Beijing	1.6239	Guangdong	1.4553
2009	Taiwan	2.4580	Hong Kong	1.8298	Beijing	1.6749	Jiangsu	1.6436
2010	Taiwan	2.2788	Hong Kong	1.7550	Beijing	1.7239	Jiangsu	1.6836
2011	Taiwan	2.2801	Beijing	1.7820	Jiangsu	1.7620	Hong Kong	1.7402
2012	Taiwan	2.2121	Jiangsu	1.8222	Guangdong	1.6910	Beijing	1.6866

(Continued)

[16]Result shown in Table 2.6 is based on 2000–2012 data, collated and analysed in 2015.

Table 2.6: (Continued)

	5		6		7		8	
Rank	**Economy**	**Score**	**Economy**	**Score**	**Economy**	**Score**	**Economy**	**Score**
2000	Shanghai	0.6295	Jiangsu	0.6096	Shandong	0.5609	Xinjiang	0.5210
2001	Jiangsu	0.7947	Beijing	0.7786	Shanghai	0.7344	Shandong	0.6658
2002	Shanghai	0.9080	Beijing	0.8812	Jiangsu	0.8622	Shandong	0.6501
2003	Shanghai	1.1927	Jiangsu	1.0469	Beijing	1.0036	Shandong	0.6807
2004	Shanghai	1.2006	Jiangsu	1.1189	Zhejiang	1.0916	Shandong	0.6569
2005	Guangdong	1.1498	Zhejiang	1.0712	Jiangsu	1.0577	Shandong	0.6869
2006	Shanghai	1.2294	Jiangsu	1.1913	Zhejiang	1.0782	Shandong	0.7522
2007	Shanghai	1.3469	Jiangsu	1.3251	Zhejiang	1.1699	Shandong	0.7922
2008	Jiangsu	1.4059	Shanghai	1.3563	Zhejiang	1.1707	Shandong	0.8227
2009	Guangdong	1.5940	Shanghai	1.4594	Zhejiang	1.2844	Shandong	0.7988
2010	Guangdong	1.5966	Shanghai	1.3956	Zhejiang	1.3286	Macau	1.0717
2011	Guangdong	1.6625	Shanghai	1.4115	Zhejiang	1.3292	Macau	0.9892
2012	Hong Kong	1.6646	Shanghai	1.4031	Zhejiang	1.3717	Macau	1.1074

	9		10		11		12	
Rank	Economy	Score	Economy	Score	Economy	Score	Economy	Score
2000	Beijing	0.4336	Liaoning	0.4307	Ningxia	0.0879	Yunnan	0.0713
2001	Liaoning	0.3440	Henan	0.1849	Fujian	0.0790	Xinjiang	0.0764
2002	Liaoning	0.2894	Henan	0.1647	Fujian	0.1618	Macau	0.0976
2003	Macau	0.2797	Liaoning	0.1635	Fujian	0.0780	Henan	0.0281
2004	Macau	0.6377	Fujian	0.0908	Liaoning	0.0471	Henan	–0.1284
2005	Macau	0.5245	Fujian	0.0799	Liaoning	0.0341	Xinjiang	–0.1758
2006	Macau	0.4406	Fujian	0.0727	Liaoning	–0.0430	Xinjiang	–0.0509
2007	Macau	0.6435	Fujian	0.0827	Xinjiang	–0.1150	Hebei	–0.1176
2008	Macau	0.7283	Fujian	0.0616	Xinjiang	–0.1009	Hebei	–0.1205
2009	Macau	0.6417	Fujian	0.0672	Hebei	–0.1292	Liaoning	–0.1405
2010	Shandong	0.7210	Fujian	–0.0216	Xinjiang	–0.1177	Hebei	–0.1214
2011	Shandong	0.7167	Fujian	–0.0052	Liaoning	–0.1310	Xinjiang	–0.1331
2012	Shandong	0.7651	Fujian	–0.0487	Liaoning	–0.0705	Sichuan	–0.1431

(*Continued*)

Table 2.6: *(Continued)*

	13		14		15		16	
Rank	**Economy**	**Score**	**Economy**	**Score**	**Economy**	**Score**	**Economy**	**Score**
2000	Macau	0.0675	Henan	0.0242	Fujian	−0.0158	Heilongjiang	−0.1568
2001	Ningxia	0.0688	Yunnan	0.0671	Macau	−0.1353	Heilongjiang	−0.1533
2002	Yunnan	0.0166	Ningxia	−0.0004	Xinjiang	−0.0703	Chongqing	−0.0874
2003	Xinjiang	−0.0986	Heilongjiang	−0.2350	Ningxia	−0.2692	Jilin	−0.2929
2004	Xinjiang	−0.1306	Sichuan	−0.2007	Ningxia	−0.2662	Yunnan	−0.2805
2005	Hebei	−0.1957	Sichuan	−0.2015	Henan	−0.2102	Inner Mongolia	−0.2148
2006	Henan	−0.1091	Sichuan	−0.1572	Hebei	−0.1626	Inner Mongolia	−0.1812
2007	Henan	−0.1397	Liaoning	−0.1653	Sichuan	−0.2093	Inner Mongolia	−0.3163
2008	Henan	−0.1712	Liaoning	−0.1754	Hubei	−0.2366	Tianjin	−0.2745
2009	Henan	−0.2082	Hubei	−0.2140	Sichuan	−0.2586	Inner Mongolia	−0.2714
2010	Liaoning	−0.1346	Hubei	−0.2424	Inner Mongolia	−0.2588	Sichuan	−0.2755
2011	Inner Mongolia	−0.1831	Hebei	−0.1944	Sichuan	−0.2046	Anhui	−0.3234
2012	Hebei	−0.1845	Inner Mongolia	−0.2327	Xinjiang	−0.2358	Hubei	−0.2516

	17		18		19		20	
Rank	Economy	Score	Economy	Score	Economy	Score	Economy	Score
2000	Jilin	−0.1686	Shanxi	−0.1707	Inner Mongolia	−0.2003	Sichuan	−0.2285
2001	Jilin	−0.1623	Sichuan	−0.2140	Inner Mongolia	−0.3349	Gansu	−0.3866
2002	Heilongjiang	−0.0907	Jilin	−0.1508	Sichuan	−0.2248	Qinghai	−0.3030
2003	Yunnan	−0.2947	Sichuan	−0.2963	Hubei	−0.3090	Hebei	−0.3593
2004	Hubei	−0.2849	Qinghai	−0.2959	Hebei	−0.2993	Jilin	−0.4134
2005	Yunnan	−0.2269	Qinghai	−0.2543	Hubei	−0.3060	Heilongjiang	−0.3416
2006	Hubei	−0.2809	Yunnan	−0.3037	Heilongjiang	−0.3048	Tianjin	−0.3115
2007	Yunnan	−0.3202	Hubei	−0.3241	Tianjin	−0.3867	Heilongjiang	−0.3883
2008	Inner Mongolia	−0.2809	Sichuan	−0.3718	Yunnan	−0.3782	Heilongjiang	−0.4158
2009	Tianjin	−0.2743	Xinjiang	−0.2951	Yunnan	−0.3304	Heilongjiang	−0.4270
2010	Henan	−0.3138	Yunnan	−0.3306	Tianjin	−0.3396	Hainan	−0.3763
2011	Hubei	−0.3326	Henan	−0.3410	Yunnan	−0.3414	Tianjin	−0.3641
2012	Yunnan	−0.3252	Anhui	−0.3281	Henan	−0.3777	Hainan	−0.3992

(*Continued*)

Table 2.6: *(Continued)*

	21		22		23		24	
Rank	**Economy**	**Score**	**Economy**	**Score**	**Economy**	**Score**	**Economy**	**Score**
2000	Tianjin	–0.2869	Guizhou	–0.4342	Qinghai	–0.4381	Chongqing	–0.4493
2001	Tianjin	–0.3965	Hubei	–0.4444	Shanxi	–0.4574	Chongqing	–0.4822
2002	Tianjin	–0.3952	Hebei	–0.4007	Hubei	–0.4144	Guizhou	–0.4707
2003	Tianjin	–0.3938	Chongqing	–0.4225	Qinghai	–0.4743	Inner Mongolia	–0.4799
2004	Tianjin	–0.4298	Chongqing	–0.4447	Inner Mongolia	–0.4909	Hainan	–0.5794
2005	Ningxia	–0.3483	Tianjin	–0.4038	Hainan	–0.4702	Guizhou	–0.5552
2006	Qinghai	–0.3465	Hainan	–0.4175	Ningxia	–0.4661	Jilin	–0.5485
2007	Hainan	–0.4421	Qinghai	–0.5086	Jilin	–0.5201	Ningxia	–0.5309
2008	Ningxia	–0.4238	Hainan	–0.4455	Jilin	–0.5205	Qinghai	–0.5546
2009	Hainan	–0.4682	Jiangxi	–0.5134	Jilin	–0.5288	Anhui	–0.5433
2010	Heilongjiang	–0.4022	Anhui	–0.4179	Ningxia	–0.5128	Shaanxi	–0.5329
2011	Heilongjiang	–0.3814	Hainan	–0.4019	Jiangxi	–0.5183	Ningxia	–0.5271
2012	Heilongjiang	–0.4812	Shaanxi	–0.4836	Jiangxi	–0.4928	Tianjin	–0.5133

	25		26		27		28	
Rank	Economy	Score	Economy	Score	Economy	Score	Economy	Score
2000	Hebei	–0.4812	Hainan	–0.5495	Hubei	–0.5589	Gansu	–0.5760
2001	Qinghai	–0.5072	Hebei	–0.5125	Guizhou	–0.5402	Jiangxi	–0.6096
2002	Inner Mongolia	–0.5011	Shanxi	–0.5546	Gansu	–0.5929	Jiangxi	–0.7063
2003	Shanxi	–0.6337	Gansu	–0.6639	Hainan	–0.6729	Guizhou	–0.6889
2004	Guizhou	–0.6005	Gansu	–0.6286	Heilongjiang	–0.6379	Shanxi	–0.7045
2005	Jilin	–0.5816	Chongqing	–0.6170	Shanxi	–0.6459	Gansu	–0.6618
2006	Chongqing	–0.5661	Gansu	–0.5768	Guizhou	–0.6571	Jiangxi	–0.6823
2007	Jiangxi	–0.5510	Chongqing	–0.6213	Anhui	–0.6621	Guizhou	–0.6688
2008	Jiangxi	–0.5581	Anhui	–0.6052	Shanxi	–0.6278	Chongqing	–0.6283
2009	Ningxia	–0.5887	Shaanxi	–0.6012	Chongqing	–0.6375	Shanxi	–0.6749
2010	Jiangxi	–0.5762	Chongqing	–0.5796	Jilin	–0.6399	Hunan	–0.6912
2011	Chongqing	–0.6100	Jilin	–0.6121	Shaanxi	–0.6586	Hunan	–0.8383
2012	Ningxia	–0.5503	Chongqing	–0.6368	Jilin	–0.6383	Hunan	–0.7580

(*Continued*)

Table 2.6: (*Continued*)

	29		30		31		32	
Rank	**Economy**	**Score**	**Economy**	**Score**	**Economy**	**Score**	**Economy**	**Score**
2000	Jiangxi	–0.6203	Guangxi	–0.8943	Anhui	–0.9339	Hunan	–1.1563
2001	Hainan	–0.7222	Guangxi	–0.8233	Anhui	–0.8714	Hunan	–1.1460
2002	Hainan	–0.7225	Anhui	–0.8799	Hunan	–1.0891	Guangxi	–1.0944
2003	Jiangxi	–0.7463	Anhui	–0.7633	Hunan	–1.0949	Guangxi	–1.0961
2004	Jiangxi	–0.7433	Anhui	–0.8004	Hunan	–0.9903	Shaanxi	–1.0465
2005	Anhui	–0.7527	Jiangxi	–0.7822	Hunan	–1.0125	Shaanxi	–1.0407
2006	Anhui	–0.6953	Hunan	–0.9239	Shaanxi	–0.9696	Guangxi	–1.0166
2007	Gansu	–0.6976	Shanxi	–0.7371	Hunan	–0.8725	Shaanxi	–0.8902
2008	Guizhou	–0.7274	Shaanxi	–0.7687	Hunan	–0.8023	Gansu	–0.9633
2009	Qinghai	–0.6832	Hunan	–0.7334	Guizhou	–0.7563	Guangxi	–0.9267
2010	Guizhou	–0.7808	Qinghai	–0.8318	Guangxi	–0.8909	Shanxi	–0.9040
2011	Shanxi	–0.8652	Guangxi	–0.8664	Qinghai	–0.9235	Guizhou	–0.9732
2012	Guangxi	–0.8003	Shanxi	–0.8516	Guizhou	–0.9100	Qinghai	–0.9960

Rank	33		34	
	Economy	Score	Economy	Score
2000	Shaanxi	−1.1826	Tibet	−1.6877
2001	Shaanxi	−1.3089	Tibet	−1.7837
2002	Shaanxi	−1.2402	Tibet	−1.9563
2003	Shaanxi	−1.1009	Tibet	−1.2526
2004	Guangxi	−1.0470	Tibet	−1.4232
2005	Guangxi	−1.1014	Tibet	−1.6727
2006	Shanxi	−1.0172	Tibet	−1.9190
2007	Guangxi	−0.9837	Tibet	−2.0304
2008	Guangxi	−1.0025	Tibet	−2.1455
2009	Gansu	−1.0442	Tibet	−2.2034
2010	Gansu	−1.0814	Tibet	−2.1809
2011	Gansu	−1.0365	Tibet	−1.9074
2012	Gansu	−1.0033	Tibet	−2.0110

Source: Asia Competitiveness Institute

spots; Gansu, Heilongjiang, Henan and Yunnan each falls five spots; Xinjiang falls seven spots; Guizhou and Qinghai each falls nine spots; Jilin falls 10 spots; Shanxi falls 12 spots; and Ningxia falls 14 spots.

Although the score of Taiwan has decreased over the years, from 4.2982 in 2000 to 2.2121 in 2012, it still outperforms the rest of the 33 economies in Greater China. In 2012, the difference in scores between Taiwan and #2-ranked Jiangsu is 0.3899, whereas the gap in scores between Taiwan and #2-ranked Hong Kong in 2000 is 2.8493. The decrease in score gap between Taiwan and the #2-ranked economy is considerable. Hong Kong has remained as the #2-ranked economy from 2000 to 2010. Beijing and Jiangsu from mainland China have surpassed Hong Kong in 2011, occupying the second and third positions respectively in the ranking. In 2012, Hong Kong has been overtaken by another mainland economy Guangdong, thereby becoming #5-ranked.

2.2.4 ACI 2015 Financial, Businesses and Manpower Conditions Ranking and Score

Table 2.7 shows the Financial, Businesses and Manpower Conditions ranking. The top performer in this environment is Hong Kong, with a score of 2.8250. Tibet is ranked last at #34 with a score of −1.1726.

Table 2.7: ACI 2015 Financial, Businesses and Manpower Conditions Ranking and Score[17]

Rank	Economy	Score
1	Hong Kong	2.8250
2	Taiwan	2.8019
3	Macau	1.4588
4	Guangdong	1.3441
5	Beijing	1.1804
6	Jiangsu	1.0780
7	Zhejiang	0.9249
8	Shanghai	0.7162
9	Shandong	0.4820
10	Liaoning	0.2315
11	Tianjin	0.1581
12	Fujian	−0.0909
13	Hebei	−0.1270
14	Inner Mongolia	−0.2059
15	Chongqing	−0.2166

(Continued)

[17]Result shown in Table 2.7 is based on 2012 data, collated and analysed in 2015.

Table 2.7: (*Continued*)

Rank	Economy	Score
16	Jilin	–0.2179
17	Henan	–0.3237
18	Hubei	–0.3434
19	Sichuan	–0.3583
20	Hunan	–0.3925
21	Anhui	–0.4474
22	Heilongjiang	–0.4740
23	Shaanxi	–0.5122
24	Xinjiang	–0.5199
25	Shanxi	–0.6029
26	Jiangxi	–0.6430
27	Hainan	–0.7900
28	Guangxi	–0.8160
29	Yunnan	–0.8813
30	Ningxia	–0.8820
31	Guizhou	–1.0378
32	Qinghai	–1.0613
33	Gansu	–1.0842
34	Tibet	–1.1726

Source: Asia Competitiveness Institute

The average performing economy (with a standardised score of 0) is situated between the #11-ranked Tianjin (0.1581) and the #12-ranked Fujian (–0.0909). While the median score is somewhere in between the #17-ranked Henan (–0.3237) and the #18-ranked Hubei (–0.3434). The median score for Financial, Businesses and Manpower Conditions ranking is hence –0.3336, which is lower than the average score (0).

The largest "jump" in score is between the #2-ranked Taiwan (2.8019) and the #3-ranked Macau (1.4588). Their difference in scores is 1.3431. The gaps between other consecutively ranked economies appear quite moderate in size. A few relatively larger gaps in score include those between #9-ranked Shandong (0.4820) and #10-ranked Liaoning (0.2315) as well as between #8-ranked Shanghai (0.7162) and #9-ranked Shandong (0.4820). Their gaps in scores are 0.2505 and 0.2342, respectively.

Figure 2.5 shows the Financial, Businesses and Manpower Conditions Ranking Map. The colours, as representative of each economy's competitiveness, are almost the same as that in Figure 2.2 (Overall Competitiveness Ranking Map) except for two economies. Chongqing and Shanxi are in red and yellow respectively in Figure 2.2, but their colours are switched in Figure 2.5.

Figure 2.5: ACI 2015 Financial, Businesses and Manpower Conditions Ranking Map[18]

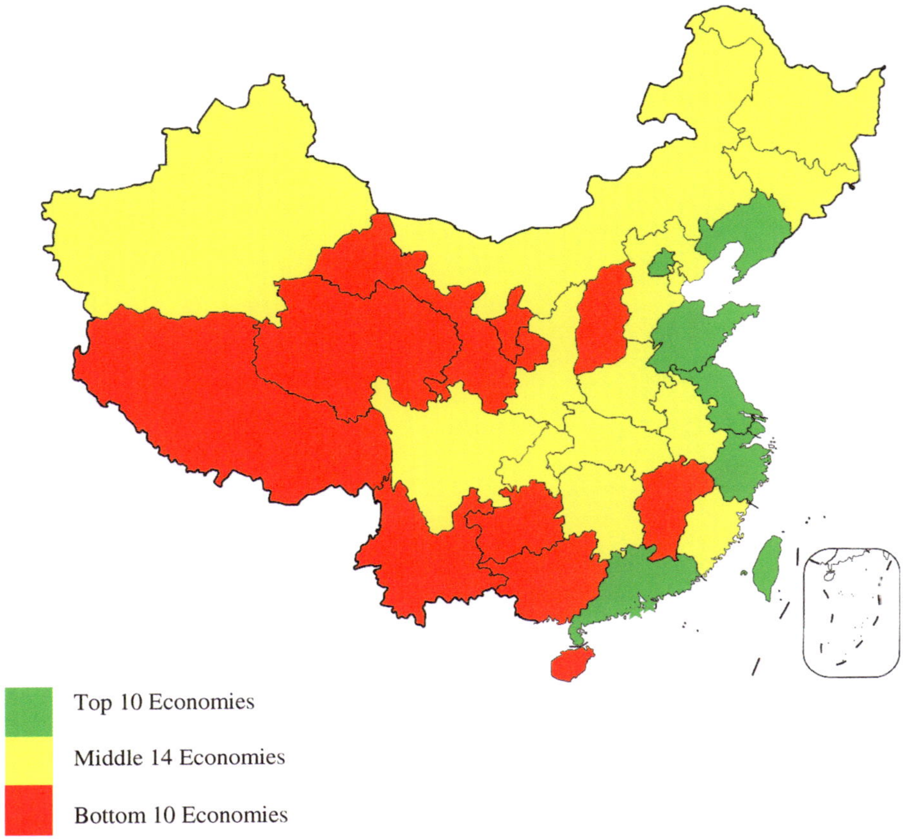

Source: Asia Competitiveness Institute edited from a map provided at http://bbs.eduol.cn/file.php?FileName=20128151347193336.jpg

The rankings and scores of the 34 Greater China economies in Financial, Businesses and Manpower Conditions are shown in Table 2.8. There are five economies whose 2012 rankings have stayed the same as in 2000, including Beijing, Fujian, Guangdong, Liaoning and Macau. There are 12 economies that have improved their standing in Financial, Businesses and Manpower Conditions over the past 13 years. Guizhou, Hainan and Hong Kong each improves one spot; Anhui, Jiangsu, Tianjin and Zhejiang each improves two spots; Hunan, Jilin and Ningxia each improves four spots; Inner Mongolia improves five spots; and Chongqing improves 15 spots. The other 17 economies have seen a decline in their respective rankings over the years. Jiangxi, Taiwan and Tibet each falls one spot; Gansu, Guangxi, Hebei, Henan, Hubei, Shandong, Shanghai, Sichuan, Xinjiang and Yunan each falls two spots; Qinghai falls three spots; Shanxi falls four spots; Shaanxi falls five spots; and Heilongjiang falls eight spots.

[18]Result shown in Figure 2.5 is based on 2012 data, collated and analysed in 2015.

Table 2.8: ACI 2015 Financial, Businesses and Manpower Conditions Ranking and Score: Evolution over Time[19]

	1		2		3		4	
Rank	**Economy**	**Score**	**Economy**	**Score**	**Economy**	**Score**	**Economy**	**Score**
Financial, Businesses and Manpower Conditions								
2000	Taiwan	4.0801	Hong Kong	3.4520	Macau	0.7754	Guangdong	0.5895
2001	Taiwan	3.8266	Hong Kong	3.6474	Macau	0.7585	Guangdong	0.7071
2002	Taiwan	3.7799	Hong Kong	3.5014	Macau	0.8714	Guangdong	0.7971
2003	Hong Kong	3.6347	Taiwan	3.5610	Macau	1.0108	Guangdong	0.9416
2004	Taiwan	3.7135	Hong Kong	3.3375	Macau	1.1230	Guangdong	0.9022
2005	Taiwan	3.7777	Hong Kong	3.0946	Macau	1.1801	Guangdong	0.9933
2006	Taiwan	3.5190	Hong Kong	3.2272	Macau	1.2506	Guangdong	1.0850
2007	Hong Kong	3.3450	Taiwan	3.2738	Macau	1.2945	Guangdong	1.1268
2008	Hong Kong	3.2791	Taiwan	3.1255	Guangdong	1.2338	Macau	1.2167
2009	Hong Kong	3.1859	Taiwan	2.8917	Guangdong	1.3391	Macau	1.1203
2010	Hong Kong	3.0478	Taiwan	2.8968	Guangdong	1.3877	Macau	1.2070
2011	Hong Kong	2.9804	Taiwan	2.3953	Macau	1.5091	Shanghai	1.4797
2012	Hong Kong	2.8250	Taiwan	2.8019	Macau	1.4588	Guangdong	1.3441

(Continued)

[19]Result shown in Table 2.8 is based on 2000–2012 data, collated and analysed in 2015.

Table 2.8: *(Continued)*

	5		6		7		8	
Rank	**Economy**	**Score**	**Economy**	**Score**	**Economy**	**Score**	**Economy**	**Score**
2000	Beijing	0.2916	Shanghai	0.2142	Shandong	0.0392	Jiangsu	0.0391
2001	Beijing	0.3888	Shanghai	0.2444	Zhejiang	0.1223	Jiangsu	0.0975
2002	Hunan	0.5380	Beijing	0.4041	Shanghai	0.2917	Zhejiang	0.1714
2003	Beijing	0.5107	Jiangsu	0.3650	Shanghai	0.3578	Zhejiang	0.3232
2004	Beijing	0.5451	Jiangsu	0.4319	Hainan	0.3693	Shanghai	0.3652
2005	Beijing	0.6171	Jiangsu	0.5496	Zhejiang	0.3946	Shanghai	0.3930
2006	Beijing	0.7687	Jiangsu	0.6748	Zhejiang	0.5380	Shanghai	0.4537
2007	Beijing	0.7871	Jiangsu	0.7446	Zhejiang	0.6135	Shanghai	0.5263
2008	Beijing	0.8869	Jiangsu	0.8513	Zhejiang	0.6601	Shanghai	0.6579
2009	Beijing	1.0221	Jiangsu	0.9956	Zhejiang	0.7768	Shanghai	0.7241
2010	Jiangsu	1.0561	Beijing	1.0153	Zhejiang	0.8144	Shanghai	0.7298
2011	Guangdong	1.3321	Beijing	1.1439	Jiangsu	1.0090	Zhejiang	0.8158
2012	Beijing	1.1804	Jiangsu	1.0780	Zhejiang	0.9249	Shanghai	0.7162

	9		10		11		12	
Rank	Economy	Score	Economy	Score	Economy	Score	Economy	Score
2000	Zhejiang	0.0039	Liaoning	–0.0660	Hebei	–0.0903	Fujian	–0.1276
2001	Shandong	0.0913	Liaoning	0.0516	Tianjin	–0.1316	Hebei	–0.1424
2002	Jiangsu	0.1473	Shandong	0.1173	Liaoning	–0.1190	Tianjin	–0.1327
2003	Shandong	0.2226	Liaoning	–0.1050	Tianjin	–0.1143	Hebei	–0.2278
2004	Zhejiang	0.3067	Shandong	0.2344	Tianjin	–0.1360	Liaoning	–0.1732
2005	Shandong	0.3896	Hainan	0.3203	Tianjin	–0.1000	Liaoning	–0.1438
2006	Shandong	0.4479	Liaoning	–0.0570	Tianjin	–0.0787	Fujian	–0.1878
2007	Shandong	0.4314	Liaoning	0.0699	Tianjin	–0.0786	Fujian	–0.1764
2008	Shandong	0.4215	Liaoning	0.1793	Tianjin	0.0571	Fujian	–0.1790
2009	Shandong	0.4913	Liaoning	0.2852	Tianjin	0.2216	Fujian	–0.1596
2010	Shandong	0.4599	Liaoning	0.2897	Tianjin	0.2322	Fujian	–0.1509
2011	Shandong	0.4908	Liaoning	0.1607	Tianjin	0.0462	Fujian	–0.1477
2012	Shandong	0.4820	Liaoning	0.2315	Tianjin	0.1581	Fujian	–0.0909

(Continued)

Table 2.8: *(Continued)*

	13		14		15		16	
Rank	**Economy**	**Score**	**Economy**	**Score**	**Economy**	**Score**	**Economy**	**Score**
2000	Tianjin	−0.1533	Heilongjiang	−0.2032	Henan	−0.2153	Hubei	−0.2401
2001	Fujian	−0.2054	Henan	−0.2607	Hubei	−0.2617	Heilongjiang	−0.2782
2002	Hebei	−0.1942	Fujian	−0.2546	Hubei	−0.2705	Heilongjiang	−0.3000
2003	Fujian	−0.2293	Heilongjiang	−0.2421	Hubei	−0.2671	Xinjiang	−0.3052
2004	Fujian	−0.2312	Hebei	−0.2732	Hubei	−0.2923	Xinjiang	−0.3305
2005	Fujian	−0.2307	Hebei	−0.2763	Henan	−0.3719	Xinjiang	−0.3743
2006	Hebei	−0.2270	Henan	−0.3403	Hubei	−0.3532	Inner Mongolia	−0.3724
2007	Hebei	−0.2460	Inner Mongolia	−0.3384	Henan	−0.3551	Hubei	−0.3721
2008	Hebei	−0.2355	Inner Mongolia	−0.2754	Henan	−0.3169	Hubei	−0.3342
2009	Inner Mongolia	−0.1986	Hebei	−0.2021	Henan	−0.2700	Hubei	−0.2789
2010	Hebei	−0.2043	Inner Mongolia	−0.2090	Hubei	−0.2347	Henan	−0.3148
2011	Hebei	−0.1692	Inner Mongolia	−0.1931	Chongqing	−0.2407	Jilin	−0.2505
2012	Hebei	−0.1270	Inner Mongolia	−0.2059	Chongqing	−0.2166	Jilin	−0.2179

	17		18		19		20	
Rank	**Economy**	**Score**	**Economy**	**Score**	**Economy**	**Score**	**Economy**	**Score**
2000	Sichuan	–0.3142	Shaanxi	–0.3396	Inner Mongolia	–0.3476	Jilin	–0.3504
2001	Xinjiang	–0.3056	Sichuan	–0.3159	Jilin	–0.3198	Inner Mongolia	–0.3596
2002	Sichuan	–0.3276	Henan	–0.3325	Xinjiang	–0.3433	Jilin	–0.3920
2003	Sichuan	–0.3157	Henan	–0.3248	Inner Mongolia	–0.4409	Anhui	–0.4436
2004	Heilongjiang	–0.3305	Henan	–0.3534	Sichuan	–0.3703	Jilin	–0.4683
2005	Heilongjiang	–0.3797	Hubei	–0.3946	Inner Mongolia	–0.4191	Sichuan	–0.4307
2006	Xinjiang	–0.3819	Heilongjiang	–0.3907	Sichuan	–0.4078	Jilin	–0.4489
2007	Jilin	–0.3818	Sichuan	–0.4026	Xinjiang	–0.4209	Heilongjiang	–0.4284
2008	Jilin	–0.3523	Xinjiang	–0.3905	Sichuan	–0.4544	Heilongjiang	–0.4655
2009	Jilin	–0.3249	Sichuan	–0.3839	Anhui	–0.4622	Xinjiang	–0.4736
2010	Jilin	–0.3465	Sichuan	–0.3845	Xinjiang	–0.3879	Anhui	–0.3984
2011	Henan	–0.3191	Hubei	–0.3314	Sichuan	–0.4188	Anhui	–0.4559
2012	Henan	–0.3237	Hubei	–0.3434	Sichuan	–0.3583	Hunan	–0.3925

(Continued)

Table 2.8: *(Continued)*

	21		22		23		24	
Rank	**Economy**	**Score**	**Economy**	**Score**	**Economy**	**Score**	**Economy**	**Score**
2000	Shanxi	−0.3813	Xinjiang	−0.3875	Anhui	−0.4113	Hunan	−0.4123
2001	Shaanxi	−0.3954	Hunan	−0.4086	Shanxi	−0.4366	Anhui	−0.4431
2002	Inner Mongolia	−0.4182	Shaanxi	−0.4395	Anhui	−0.4541	Jiangxi	−0.4897
2003	Shaanxi	−0.4486	Hunan	−0.4686	Shanxi	−0.4864	Jiangxi	−0.4918
2004	Anhui	−0.4879	Inner Mongolia	−0.4934	Jiangxi	−0.5035	Shaanxi	−0.5055
2005	Jilin	−0.4603	Shanxi	−0.4992	Jiangxi	−0.5013	Hunan	−0.5102
2006	Jiangxi	−0.4914	Anhui	−0.5068	Shanxi	−0.5083	Hunan	−0.5229
2007	Anhui	−0.4936	Chongqing	−0.4963	Shanxi	−0.5176	Jiangxi	−0.5282
2008	Anhui	−0.5007	Chongqing	−0.5100	Shaanxi	−0.5387	Shanxi	−0.5472
2009	Chongqing	−0.4785	Hunan	−0.5741	Shaanxi	−0.5767	Heilongjiang	−0.5818
2010	Chongqing	−0.4943	Jiangxi	−0.5470	Heilongjiang	−0.5527	Shanxi	−0.5576
2011	Heilongjiang	−0.4628	Xinjiang	−0.4893	Hunan	−0.5424	Shanxi	−0.5734
2012	Anhui	−0.4474	Heilongjiang	−0.4740	Shaanxi	−0.5122	Xinjiang	−0.5199

	25		26		27		28	
Rank	**Economy**	**Score**	**Economy**	**Score**	**Economy**	**Score**	**Economy**	**Score**
2000	Jiangxi	–0.4232	Guangxi	–0.4293	Yunnan	–0.4337	Hainan	–0.4706
2001	Jiangxi	–0.4610	Yunnan	–0.4622	Guangxi	–0.4664	Hainan	–0.5135
2002	Guangxi	–0.5113	Shanxi	–0.5358	Yunnan	–0.5420	Hainan	–0.5465
2003	Jilin	–0.4988	Chongqing	–0.4997	Guangxi	–0.5256	Yunnan	–0.5312
2004	Hunan	–0.5261	Shanxi	–0.5287	Chongqing	–0.5534	Yunnan	–0.5792
2005	Anhui	–0.5413	Chongqing	–0.5541	Shaanxi	–0.5784	Yunnan	–0.6009
2006	Chongqing	–0.5280	Shaanxi	–0.5308	Yunnan	–0.6355	Guangxi	–0.6474
2007	Shaanxi	–0.5611	Hunan	–0.5670	Yunnan	–0.6481	Guangxi	–0.6510
2008	Jiangxi	–0.5544	Hunan	–0.5894	Yunnan	–0.6968	Guangxi	–0.7005
2009	Jiangxi	–0.5869	Shanxi	–0.6024	Ningxia	–0.7636	Yunnan	–0.7656
2010	Hunan	–0.5815	Shaanxi	–0.5917	Hainan	–0.7514	Ningxia	–0.7764
2011	Shaanxi	–0.6120	Jiangxi	–0.6190	Hainan	–0.7664	Guangxi	–0.8057
2012	Shanxi	–0.6029	Jiangxi	–0.6430	Hainan	–0.7900	Guangxi	–0.8160

(Continued)

Table 2.8: *(Continued)*

	29		30		31		32	
Rank	Economy	Score	Economy	Score	Economy	Score	Economy	Score
2000	Qinghai	–0.5368	Chongqing	–0.5395	Gansu	–0.6118	Guizhou	–0.6240
2001	Chongqing	–0.5223	Gansu	–0.5811	Qinghai	–0.5930	Guizhou	–0.6445
2002	Chongqing	–0.5495	Qinghai	–0.6327	Gansu	–0.6693	Guizhou	–0.7015
2003	Hainan	–0.6197	Ningxia	–0.6360	Qinghai	–0.6505	Tibet	–0.6662
2004	Guangxi	–0.6117	Ningxia	–0.6643	Tibet	–0.7089	Qinghai	–0.7093
2005	Guangxi	–0.6355	Ningxia	–0.7015	Qinghai	–0.7068	Tibet	–0.7527
2006	Ningxia	–0.6526	Hainan	–0.6754	Gansu	–0.7367	Qinghai	–0.7426
2007	Hainan	–0.6855	Ningxia	–0.6964	Tibet	–0.7667	Gansu	–0.7673
2008	Hainan	–0.7337	Ningxia	–0.7500	Gansu	–0.8025	Tibet	–0.8490
2009	Guangxi	–0.7749	Hainan	–0.7845	Tibet	–0.9237	Gansu	–0.9278
2010	Guangxi	–0.7830	Yunnan	–0.8201	Gansu	–0.9723	Qinghai	–0.9968
2011	Yunnan	–0.8466	Ningxia	–0.8489	Gansu	–0.9910	Guizhou	–1.0792
2012	Yunnan	–0.8813	Ningxia	–0.8820	Guizhou	–1.0378	Qinghai	–1.0613

Rank	33		34	
	Economy	Score	Economy	Score
2000	Tibet	–0.6698	Ningxia	–0.7064
2001	Tibet	–0.7132	Ningxia	–0.7138
2002	Ningxia	–0.7084	Tibet	–0.7547
2003	Gansu	–0.6909	Guizhou	–0.6976
2004	Guizhou	–0.7447	Gansu	–0.7532
2005	Guizhou	–0.7687	Gansu	–0.7778
2006	Tibet	–0.7641	Guizhou	–0.7766
2007	Qinghai	–0.7906	Guizhou	–0.8434
2008	Qinghai	–0.8495	Guizhou	–0.9431
2009	Qinghai	–0.9522	Guizhou	–1.0072
2010	Tibet	–1.0286	Guizhou	–1.0525
2011	Qinghai	–1.0865	Tibet	–1.1128
2012	Gansu	–1.0842	Tibet	–1.1726

Source: Asia Competitiveness Institute

Before 2008, Hong Kong, Macau and Taiwan have always occupied the top three positions in the Financial, Businesses and Manpower Conditions ranking. From 2008 to 2010, Macau has been briefly displaced from the top three positions by a rising Guangdong. However, Macau has regained the third position in the ranking of this environment since 2011.

2.2.5 ACI 2015 Quality of Life and Infrastructure Development Ranking and Score

Table 2.9 shows the ranking of Quality of Life and Infrastructure Development. Beijing is ranked first with the highest score (2.3659) among the 34 Greater China economies, while Tibet is ranked last at a score of –2.0100.

The average performing economy (with a standardised score of 0) is somewhere between the #18-ranked Hubei (0.0005) and the #19-ranked Jilin (–0.1190). The median score, which is defined to be at the middle of a ranking, falls between the #17-ranked Heilongjiang (0.0120) and the #18-ranked Hubei (0.0005). The median score for Quality of Life and Infrastructure Development ranking is 0.0063. For this environment, the median score (0.0063) is slightly higher than the average score (0).

Table 2.9: ACI 2015 Quality of Life and Infrastructure Development Ranking and Score[20]

Rank	Economy	Score
1	Beijing	2.3659
2	Guangdong	1.6918
3	Shanghai	1.6167
4	Jiangsu	1.6156
5	Zhejiang	1.3711
6	Shandong	0.7669
7	Taiwan	0.7472
8	Hebei	0.6600
9	Tianjin	0.5720
10	Fujian	0.4571
11	Liaoning	0.3935
12	Inner Mongolia	0.3469
13	Xinjiang	0.2779
14	Macau	0.2208
15	Shanxi	0.1691

(Continued)

[20]Result shown in Table 2.9 is based on 2012 data, collated and analysed in 2015.

Table 2.9: ***(Continued)***

Rank	Economy	Score
16	Shaanxi	0.1557
17	Heilongjiang	0.0120
18	Hubei	0.0005
19	Jilin	–0.1190
20	Sichuan	–0.1651
21	Henan	–0.1942
22	Hong Kong	–0.2098
23	Anhui	–0.3881
24	Hunan	–0.5739
25	Yunnan	–0.5814
26	Hainan	–0.8991
27	Guangxi	–0.9368
28	Qinghai	−1.0626
29	Ningxia	−1.1034
30	Jiangxi	−1.1050
31	Chongqing	−1.2171
32	Gansu	−1.3518
33	Guizhou	−1.5234
34	Tibet	–2.0100

Source: Asia Competitiveness Institute

Some of the larger gaps in score include those between #1-ranked Beijing (2.3659) and #2-ranked Guangdong (1.6918), between #5-ranked Zhejiang (1.3711) and #6-ranked Shandong (0.7669), as well as between #33-ranked Guizhou (−1.5234) and #34-ranked Tibet (–2.0100). Their differences in scores are 0.6741, 0.6042 and 0.4866, respectively.

Figure 2.6 provides a snapshot of the Quality of Life and Infrastructure Development ranking. Green represents the economies that are in the top 10, yellow in the middle 14, and red in the bottom 10. The overall pattern of an increasing competitiveness going from west to east in Greater China still largely persists as shown in this map.

Table 2.10 displays the Quality of Life and Infrastructure Development ranking and score for 34 Greater China economies from 2000 to 2012. There is only one economy that has retained the same ranking in 2012 as compared to 2000, which is Chongqing. Meanwhile, 16 economies have moved up the Quality of Life and Infrastructure Development ranking over the past 13 years. Beijing, Guangdong, Guangxi, Guizhou and Inner Mongolia each rises one spot; Hebei and Shandong each rises two spots; Macau, Shanghai and Zhejiang each rises four spots; Hainan rises six spots; Anhui and Tianjin each rises seven spots; Jiangsu rises 10 spots; Shaanxi rises

Figure 2.6: ACI 2015 Quality of Life and Infrastructure Development Ranking Map[21]

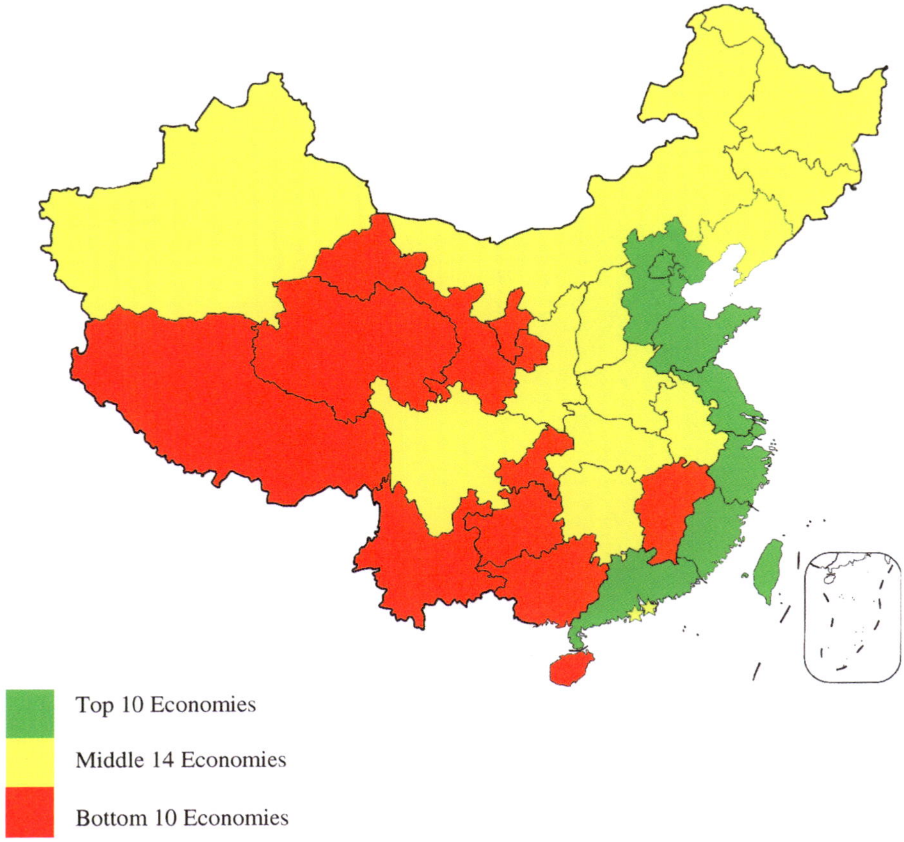

Source: Asia Competitiveness Institute edited from a map provided at http://bbs.eduol.cn/file.php?FileName=20128151347193336.jpg

11 spots; and Fujian rises 12 spots. From 2000 to 2012, the ranking of 17 economies have dropped. Henan, Hubei, Sichuan and Tibet each falls one spot; Xinjiang and Yunnan each falls two spots; Gansu, Hunan and Shanxi each falls three spots; Jiangxi, Jilin, Ningxia and Qinghai each falls four spots; Taiwan falls six spots; Liaoning falls seven spots; Heilongjiang falls 12 spots; and Hong Kong falls 16 spots.

Unlike the rankings of other environments, we do not see Hong Kong, Macau or Taiwan occupying the top three positions for the Quality of Life and Infrastructure Development ranking from 2008 onwards. Instead, Beijing, Guangdong and Shanghai have been performing well in this particular environment.

---

[21] Result shown in Figure 2.6 is based on 2012 data, collated and analysed in 2015.

Table 2.10: ACI 2015 Quality of Life and Infrastructure Development Ranking and Score: Evolution over Time[22]

	1		2		3		4	
Rank	**Economy**	**Score**	**Economy**	**Score**	**Economy**	**Score**	**Economy**	**Score**
Quality of Life and Infrastructure Development								
2000	Taiwan	3.4793	Beijing	1.8471	Guangdong	1.2374	Liaoning	1.0897
2001	Taiwan	3.3756	Beijing	1.8314	Guangdong	1.1459	Liaoning	0.9992
2002	Taiwan	3.2255	Beijing	1.6116	Guangdong	1.5209	Liaoning	1.1360
2003	Taiwan	2.9596	Beijing	1.8688	Guangdong	1.4147	Liaoning	0.9773
2004	Taiwan	2.7360	Beijing	1.7067	Guangdong	1.4464	Hebei	1.0249
2005	Taiwan	2.3459	Beijing	1.7228	Guangdong	1.7048	Hebei	0.9995
2006	Taiwan	2.0445	Guangdong	1.8396	Beijing	1.7936	Shandong	1.1984
2007	Beijing	2.1910	Taiwan	1.8205	Guangdong	1.6460	Shanghai	1.4381
2008	Beijing	2.3179	Guangdong	1.5721	Shanghai	1.5045	Taiwan	1.4612
2009	Beijing	2.3404	Guangdong	1.4836	Shanghai	1.4658	Taiwan	1.3055
2010	Beijing	2.3387	Shanghai	1.6838	Guangdong	1.5866	Jiangsu	1.3831
2011	Beijing	2.2014	Shanghai	1.8245	Guangdong	1.5894	Jiangsu	1.5773
2012	Beijing	2.3659	Guangdong	1.6918	Shanghai	1.6167	Jiangsu	1.6156

(Continued)

[22]Result shown in Table 2.10 is based on 2000–2012 data, collated and analysed in 2015.

Table 2.10: (*Continued*)

	5		6		7		8	
Rank	**Economy**	**Score**	**Economy**	**Score**	**Economy**	**Score**	**Economy**	**Score**
2000	Heilongjiang	0.7324	Hong Kong	0.6611	Shanghai	0.5191	Shandong	0.4304
2001	Hong Kong	0.8303	Shanghai	0.6925	Heilongjiang	0.6725	Tianjin	0.5465
2002	Hong Kong	0.7836	Shanghai	0.7564	Heilongjiang	0.7089	Hebei	0.6315
2003	Hong Kong	0.8447	Hebei	0.7906	Heilongjiang	0.7269	Shanghai	0.7215
2004	Liaoning	0.9237	Shandong	0.8032	Zhejiang	0.7873	Hong Kong	0.7502
2005	Shanghai	0.9582	Shandong	0.8896	Zhejiang	0.8797	Jiangsu	0.8406
2006	Zhejiang	1.0492	Jiangsu	0.9866	Shanghai	0.8212	Hebei	0.7262
2007	Jiangsu	1.0280	Hebei	0.9609	Shandong	0.9394	Zhejiang	0.9279
2008	Shandong	1.1083	Jiangsu	1.0982	Zhejiang	1.0839	Hebei	1.0063
2009	Jiangsu	1.2323	Shandong	1.0972	Zhejiang	1.0830	Hebei	0.9717
2010	Zhejiang	1.1834	Hebei	0.9308	Shandong	0.9079	Tianjin	0.8722
2011	Zhejiang	1.2830	Taiwan	0.9639	Hebei	0.8032	Shandong	0.8031
2012	Zhejiang	1.3711	Shandong	0.7669	Taiwan	0.7472	Hebei	0.6600

Rank	9		10		11		12	
	Economy	Score	Economy	Score	Economy	Score	Economy	Score
2000	Zhejiang	0.3958	Hebei	0.3365	Xinjiang	0.3207	Shanxi	0.3128
2001	Shandong	0.4864	Hebei	0.4531	Zhejiang	0.4250	Shanxi	0.2978
2002	Shandong	0.6027	Tianjin	0.4469	Jiangsu	0.3393	Zhejiang	0.3248
2003	Shandong	0.6299	Tianjin	0.4759	Zhejiang	0.4505	Jiangsu	0.3820
2004	Heilongjiang	0.6972	Shanghai	0.6067	Jiangsu	0.5553	Macau	0.5293
2005	Liaoning	0.7288	Macau	0.6315	Tianjin	0.5955	Hong Kong	0.5561
2006	Liaoning	0.7078	Macau	0.7037	Tianjin	0.6874	Hong Kong	0.5791
2007	Macau	0.6420	Tianjin	0.6059	Liaoning	0.5506	Xinjiang	0.4292
2008	Tianjin	0.6526	Liaoning	0.4971	Macau	0.4458	Hong Kong	0.3318
2009	Tianjin	0.7259	Liaoning	0.5724	Inner Mongolia	0.3592	Hong Kong	0.3452
2010	Taiwan	0.8416	Fujian	0.4152	Inner Mongolia	0.4137	Liaoning	0.4123
2011	Tianjin	0.6070	Liaoning	0.4573	Fujian	0.4078	Inner Mongolia	0.3825
2012	Tianjin	0.5720	Fujian	0.4571	Liaoning	0.3935	Inner Mongolia	0.3469

(Continued)

Table 2.10: *(Continued)*

	13		14		15		16	
Rank	Economy	Score	Economy	Score	Economy	Score	Economy	Score
2000	Inner Mongolia	0.2491	Jiangsu	0.2072	Jilin	0.1891	Tianjin	0.1773
2001	Xinjiang	0.2864	Jiangsu	0.1945	Jilin	0.1633	Inner Mongolia	0.1379
2002	Shanxi	0.2706	Xinjiang	0.2533	Jilin	0.2400	Macau	0.1953
2003	Xinjiang	0.3404	Shanxi	0.2597	Macau	0.2157	Jilin	0.0915
2004	Tianjin	0.5163	Xinjiang	0.2968	Shanxi	0.2457	Inner Mongolia	0.0784
2005	Heilongjiang	0.5474	Shanxi	0.2397	Xinjiang	0.1968	Inner Mongolia	0.0781
2006	Heilongjiang	0.4362	Xinjiang	0.1476	Shanxi	0.1375	Henan	0.1075
2007	Hong Kong	0.4239	Heilongjiang	0.3422	Henan	0.1591	Shanxi	0.1282
2008	Heilongjiang	0.3127	Xinjiang	0.2343	Shanxi	0.1911	Inner Mongolia	0.1686
2009	Heilongjiang	0.3271	Xinjiang	0.2136	Macau	0.1488	Shanxi	0.1224
2010	Xinjiang	0.2777	Heilongjiang	0.2719	Macau	0.2061	Shanxi	0.1283
2011	Xinjiang	0.3720	Heilongjiang	0.2170	Macau	0.1864	Shanxi	0.1082
2012	Xinjiang	0.2779	Macau	0.2208	Shanxi	0.1691	Shaanxi	0.1557

	17		18		19		20	
Rank	**Economy**	**Score**	**Economy**	**Score**	**Economy**	**Score**	**Economy**	**Score**
2000	Hubei	0.1650	Macau	–0.0288	Sichuan	–0.0498	Henan	–0.1688
2001	Hubei	0.0363	Macau	–0.0050	Sichuan	–0.0454	Henan	–0.1156
2002	Inner Mongolia	0.0094	Sichuan	–0.0419	Fujian	–0.1289	Henan	–0.1382
2003	Sichuan	0.0553	Inner Mongolia	–0.0066	Hubei	–0.0900	Fujian	–0.1357
2004	Jilin	–0.0069	Sichuan	–0.0738	Fujian	–0.1513	Henan	–0.1664
2005	Jilin	0.0364	Henan	–0.0495	Fujian	–0.0668	Hubei	–0.1505
2006	Inner Mongolia	0.0804	Jilin	0.0202	Fujian	–0.1676	Shaanxi	–0.1874
2007	Inner Mongolia	0.0154	Jilin	–0.1536	Hunan	–0.2248	Hubei	–0.2590
2008	Henan	0.1099	Fujian	–0.0952	Hubei	–0.1928	Jilin	–0.2237
2009	Fujian	0.0521	Henan	0.0442	Jilin	0.0083	Shaanxi	–0.1203
2010	Shaanxi	0.0219	Hong Kong	0.0178	Hubei	–0.1435	Henan	–0.1490
2011	Hong Kong	0.0400	Shaanxi	–0.0511	Hubei	–0.1032	Sichuan	–0.2542
2012	Heilongjiang	0.0120	Hubei	0.0005	Jilin	–0.1190	Sichuan	–0.1651

(Continued)

Table 2.10: *(Continued)*

	21		22		23		24	
Rank	**Economy**	**Score**	**Economy**	**Score**	**Economy**	**Score**	**Economy**	**Score**
2000	Hunan	−0.2012	Fujian	−0.2551	Yunnan	−0.4643	Qinghai	−0.5562
2001	Hunan	−0.2312	Fujian	−0.3084	Yunnan	−0.4524	Shaanxi	−0.5500
2002	Hunan	−0.3157	Hubei	−0.4514	Yunnan	−0.5500	Ningxia	−0.5780
2003	Henan	−0.1885	Hunan	−0.3602	Ningxia	−0.5507	Shaanxi	−0.5589
2004	Hubei	−0.1843	Hunan	−0.3611	Shaanxi	−0.4469	Ningxia	−0.6224
2005	Hunan	−0.3226	Shaanxi	−0.3520	Sichuan	−0.5191	Yunnan	−0.6886
2006	Hubei	−0.1909	Hunan	−0.2684	Sichuan	−0.4131	Yunnan	−0.8093
2007	Shaanxi	−0.3345	Fujian	−0.3634	Sichuan	−0.3929	Guangxi	−0.7638
2008	Shaanxi	−0.2243	Hunan	−0.3796	Sichuan	−0.5065	Guangxi	−0.6608
2009	Hubei	−0.1701	Sichuan	−0.4270	Hunan	−0.4283	Guangxi	−0.7383
2010	Sichuan	−0.2491	Jilin	−0.3020	Hunan	−0.3379	Anhui	−0.7647
2011	Jilin	−0.2667	Henan	−0.3658	Hunan	−0.4578	Anhui	−0.6367
2012	Henan	−0.1942	Hong Kong	−0.2098	Anhui	−0.3881	Hunan	−0.5739

	25		26		27		28	
Rank	Economy	Score	Economy	Score	Economy	Score	Economy	Score
2000	Ningxia	–0.5721	Jiangxi	–0.5862	Shaanxi	–0.6288	Guangxi	–0.9653
2001	Ningxia	–0.5838	Jiangxi	–0.6403	Qinghai	–0.7327	Guangxi	–0.9864
2002	Jiangxi	–0.6412	Qinghai	–0.6966	Shaanxi	–0.6973	Chongqing	–0.8298
2003	Yunnan	–0.6582	Jiangxi	–0.7109	Qinghai	–0.7163	Anhui	–0.9685
2004	Jiangxi	–0.7353	Yunnan	–0.7727	Qinghai	–0.8557	Guangxi	–0.9364
2005	Jiangxi	–0.7021	Ningxia	–0.7080	Qinghai	–0.8852	Guangxi	–0.8929
2006	Jiangxi	–0.8309	Guangxi	–0.8321	Ningxia	–1.0101	Qinghai	–1.0361
2007	Jiangxi	–0.8711	Yunnan	–0.8809	Qinghai	–0.9390	Ningxia	–0.9743
2008	Yunnan	–0.7284	Jiangxi	–0.8113	Qinghai	–0.9417	Ningxia	–0.9806
2009	Jiangxi	–0.7948	Anhui	–0.9106	Yunnan	–0.9550	Ningxia	–1.0222
2010	Yunnan	–0.8325	Guangxi	–0.8731	Jiangxi	–1.0310	Hainan	–1.0469
2011	Guangxi	–0.6579	Yunnan	–0.8342	Jiangxi	–0.9776	Hainan	–1.0073
2012	Yunnan	–0.5814	Hainan	–0.8991	Guangxi	–0.9368	Qinghai	–1.0626

(*Continued*)

Table 2.10: (*Continued*)

	29		30		31		32	
Rank	**Economy**	**Score**	**Economy**	**Score**	**Economy**	**Score**	**Economy**	**Score**
2000	Gansu	−1.0005	Anhui	−1.1568	Chongqing	−1.2179	Hainan	−1.2193
2001	Chongqing	−1.0258	Gansu	−1.0855	Anhui	−1.0884	Hainan	−1.2138
2002	Guangxi	−1.0547	Anhui	−1.0854	Hainan	−1.2103	Gansu	−1.2742
2003	Chongqing	−1.0432	Guangxi	−1.1400	Hainan	−1.1970	Gansu	−1.3400
2004	Anhui	−0.9769	Hainan	−1.1504	Gansu	−1.3665	Chongqing	−1.3959
2005	Chongqing	−1.1565	Anhui	−1.1723	Hainan	−1.3130	Gansu	−1.3884
2006	Anhui	−1.1021	Chongqing	−1.1100	Hainan	−1.2938	Gansu	−1.4027
2007	Anhui	−1.0488	Chongqing	−1.1626	Hainan	−1.2097	Tibet	−1.3586
2008	Anhui	−1.0594	Chongqing	−1.2753	Hainan	−1.3201	Tibet	−1.3815
2009	Qinghai	−1.0457	Hainan	−1.1852	Chongqing	−1.2105	Tibet	−1.5209
2010	Ningxia	−1.0800	Qinghai	−1.1426	Chongqing	−1.2460	Tibet	−1.3256
2011	Qinghai	−1.0326	Ningxia	−1.0386	Chongqing	−1.2891	Gansu	−1.3906
2012	Ningxia	−1.1034	Jiangxi	−1.1050	Chongqing	−1.2171	Gansu	−1.3518

	33		34	
Rank	**Economy**	**Score**	**Economy**	**Score**
2000	Tibet	−1.2386	Guizhou	−2.0401
2001	Tibet	−1.4706	Guizhou	−2.0395
2002	Tibet	−1.3595	Guizhou	−2.0037
2003	Tibet	−1.5362	Guizhou	−2.0043
2004	Tibet	−1.4890	Guizhou	−2.0123
2005	Tibet	−1.5993	Guizhou	−1.9848
2006	Tibet	−1.5193	Guizhou	−1.8930
2007	Gansu	−1.5022	Guizhou	−1.8092
2008	Gansu	−1.5944	Guizhou	−1.7206
2009	Gansu	−1.6339	Guizhou	−1.7359
2010	Gansu	−1.6029	Guizhou	−1.7665
2011	Tibet	−1.5095	Guizhou	−1.9510
2012	Guizhou	−1.5234	Tibet	−2.0100

Source: Asia Competitiveness Institute

2.2.6 ACI 2015 What-if Competitiveness Simulation Analysis for Overall Competitiveness

All figures and tables shown in earlier sections (Tables 2.1–2.10 and Figures 2.2–2.6) reflect results before simulation analysis. In this section, we will analyse the results of What-if Competitiveness Simulation Analysis for 34 Greater China economies. Simulation analysis indicates the potential of each economy in improving their standings. It is based on improving the top 20% weakest indicators of an economy to the average, while assuming all other economies remain constant.

Table 2.11 summarises the Overall Competitiveness result before and after simulation for each economy. Economies that showed major potential for improvement include Chongqing (up 12 spots), Tibet (up 12 spots), Guangxi (up 10 spots), Hainan (up 10 spots) and Yunnan (up 10 spots). These economies tend to be ranked low before simulation analysis. The rankings

Table 2.11: ACI 2015 What-if Competitiveness Simulation Analysis on Overall Competitiveness Ranking and Score[23]

	Rank		Score	
Economy	**Before**	**After**	**Before**	**After**
Anhui	19	14	–0.3646	–0.0145
Beijing	4	2	1.6291	1.9682
Chongqing	26	14	–0.6536	–0.1712
Fujian	11	11	0.1578	0.3987
Gansu	33	27	–1.2443	–0.7296
Guangdong	1	1	2.0367	2.2772
Guangxi	29	19	–0.8529	–0.3518
Guizhou	32	25	–1.1619	–0.6190
Hainan	28	18	–0.8450	–0.3429
Hebei	13	11	0.0595	0.3122
Heilongjiang	22	14	–0.4317	–0.0257
Henan	17	11	–0.2842	0.1752
Hong Kong	5	1	1.6120	2.2027
Hubei	15	12	–0.1833	0.1053
Hunan	23	14	–0.5517	–0.0306
Inner Mongolia	16	12	–0.2043	0.1369
Jiangsu	2	1	1.9901	2.2410
Jiangxi	27	18	–0.7870	–0.3323
Jilin	21	14	–0.4076	–0.0420

(Continued)

[23]Results shown in Tables 2.11–2.15 are based on 2012 data, collated and analysed in 2015.

Table 2.11: (*Continued*)

Economy	Rank		Score	
	Before	After	Before	After
Liaoning	9	9	0.4251	0.7436
Macau	10	8	0.4167	1.1054
Ningxia	30	23	−0.9870	−0.4844
Qinghai	31	23	−1.1515	−0.5553
Shaanxi	18	14	−0.3299	−0.0222
Shandong	8	8	0.9132	1.2260
Shanghai	6	3	1.5189	1.8802
Shanxi	24	15	−0.5588	−0.1932
Sichuan	14	11	−0.1626	0.2509
Taiwan	3	1	1.7056	2.2148
Tianjin	12	9	0.0682	0.4564
Tibet	34	22	−1.6789	−0.4544
Xinjiang	20	12	−0.3816	0.0587
Yunnan	25	15	−0.6348	−0.1977
Zhejiang	7	6	1.3244	1.5929

Source: Asia Competitiveness Institute

for Fujian, Guangdong, Liaoning and Shandong remain the same after the simulation. Although the scores of all the four economies have improved after the simulation, the improvements are not sufficient to overcome the gaps between them and their better-ranked counterparts (except for Guandong which is ranked top and thus remains at first position). Other economies within Greater China have shown some improvements in ranking. Economies that have improved by one spot after the simulation include Jiangsu and Zhejiang.

It is worth noting that the rankings for Jiangsu, Taiwan and Hong Kong which are #2, #3 and #5 respectively before simulation, all rise to #1 after simulation analysis. This shows that Jiangsu, Taiwan and Hong Kong are not far from #1-ranked Guangdong or from each other in terms of Overall Competitiveness and their potential for becoming the top of the league. The key to the ranking war lies in which economy is able to improve their top 20% weakest indicators the fastest.

2.2.7 ACI 2015 What-if Competitiveness Simulation Analysis for Four Environments

Table 2.12 illustrates the 2015 What-if Competitiveness Simulation Analysis for Macroeconomic Stability. Some economies have shown promising potential to improve their rankings, including Macau (up 13 spots), Xinjiang (up nine spots), Ningxia (up eight spots), Qinghai (up eight spots)

Table 2.12: ACI 2015 What-if Competitiveness Simulation Analysis on Macroeconomic Stability Ranking and Score

	Rank		Score	
Economy	**Before**	**After**	**Before**	**After**
Anhui	16	14	–0.1488	–0.1075
Beijing	8	7	0.6305	1.1149
Chongqing	18	17	–0.2820	–0.1921
Fujian	10	10	0.2505	0.2505
Gansu	33	27	–1.0392	–0.7270
Guangdong	2	1	2.6039	2.7330
Guangxi	22	20	–0.5169	–0.4734
Guizhou	26	23	–0.7109	–0.5552
Hainan	30	23	–0.9530	–0.5841
Hebei	15	11	–0.1345	0.0780
Heilongjiang	24	21	–0.6104	–0.4994
Henan	14	12	–0.1272	0.0529
Hong Kong	4	3	1.5224	2.0549
Hubei	13	11	–0.0652	0.1165
Hunan	17	13	–0.2612	–0.0425
Inner Mongolia	25	20	–0.6436	–0.4009
Jiangsu	1	1	2.6472	2.6957
Jiangxi	23	20	–0.5920	–0.4625
Jilin	20	17	–0.4919	–0.1878
Liaoning	7	7	0.9756	0.9756
Macau	34	21	–1.2874	–0.5228
Ningxia	31	23	–1.0166	–0.5671
Qinghai	32	24	–1.0246	–0.6164
Shaanxi	19	13	–0.3474	–0.0224
Shandong	5	5	1.2727	1.2727
Shanghai	3	3	1.7310	1.9188
Shanxi	27	19	–0.7260	–0.3228
Sichuan	11	11	0.0813	0.1295
Taiwan	9	8	0.3778	0.7713
Tianjin	12	11	0.0286	0.1782
Tibet	28	23	–0.8492	–0.5873
Xinjiang	29	20	–0.8958	–0.3742
Yunnan	21	19	–0.4970	–0.2919
Zhejiang	6	5	1.0992	1.2534

Source: Asia Competitiveness Institute

and Shanxi (up eight spots). Other economies have shown some slight improvements, with Beijing, Chongqing, Guangdong, Hong Kong, Taiwan, Tianjin and Zhejiang each rising one spot. There are six economies whose ranking remain the same even after simulation. They are Fujian, Jiangsu, Liaoning, Shandong, Shanghai and Sichuan.

Unlike the case of Overall Competitiveness, whereby the #2, #3 and #5-ranked economies are able to attain the #1 position after simulation, only the #2-ranked Guangdong can become the number one after simulation. The #3-ranked Shanghai and #5-ranked Shandong retain their original rankings after the simulation, although their respective scores have increased. This is due to the large gap in scores between each economy and the economy ranked above it. Taking Shanghai as an example, it is ranked #3 with a score of 1.7310 before simulation but after simulation, its score has improved to 1.9188. Although Shanghai's score has increased after simulation, the competitiveness of Shanghai still lags behind #2-ranked Guangdong in this environment. The increase in score for Shanghai is still insufficient to surpass that of Guangdong. Similar explanations can also be applied to Shandong, Liaoning, Fujian and Sichuan.

The inability of the top ranked economies to surpass the economy ranked above them after simulation hints that the competitiveness of Shanghai, Hong Kong and Shandong are far from Jiangsu, and from each other. Among the six economies whose rankings remain the same after simulation, the scores of Fujian, Liaoning and Shandong also remain the same after simulation. Fujian has retained its #10 position in the Macroeconomic Stability ranking with a score of 0.2505. Liaoning has retained its #7 position with a score of 0.9756. Shandong has retained its #5 position with a score of 1.2727. Such phenomenon is due to the absence of indicators of this environment within the top 20% weakest indicators of Fujian, Liaoning and Shandong.

Table 2.13 shows the Government and Institutional Setting simulation results for 34 Greater China economies. Tibet, Qinghai, Gansu, Chongqing and Jilin have potential for massive improvements, with their ranking rising by 24 spots, 19 spots, 17 spots, 16 spots and 15 spots,

Table 2.13: ACI 2015 What-if Competitiveness Simulation Analysis on Government and Institutional Setting Ranking and Score

	Rank		Score	
Economy	**Before**	**After**	**Before**	**After**
Anhui	18	14	–0.3281	–0.2062
Beijing	4	4	1.6866	1.6866
Chongqing	26	10	–0.6368	–0.0093
Fujian	10	10	–0.0487	0.2197
Gansu	33	16	–1.0033	–0.2764
Guangdong	3	2	1.6910	1.8760
Guangxi	29	19	–0.8003	–0.3762
Guizhou	31	26	–0.9100	–0.6154
Hainan	20	10	–0.3992	0.2522

(Continued)

Table 2.13: (*Continued*)

	Rank		Score	
Economy	**Before**	**After**	**Before**	**After**
Hebei	13	13	–0.1845	–0.1845
Heilongjiang	21	10	–0.4812	–0.0101
Henan	19	17	–0.3777	–0.3091
Hong Kong	5	2	1.6646	1.9064
Hubei	16	12	–0.2516	–0.1207
Hunan	28	14	–0.7580	–0.2280
Inner Mongolia	14	11	–0.2327	–0.0737
Jiangsu	2	2	1.8222	2.0385
Jiangxi	23	23	–0.4928	–0.4928
Jilin	27	12	–0.6383	–0.1576
Liaoning	11	10	–0.0705	0.3626
Macau	8	7	1.1074	1.3507
Ningxia	25	12	–0.5503	–0.1152
Qinghai	32	13	–0.9960	–0.2020
Shaanxi	22	13	–0.4836	–0.1735
Shandong	9	9	0.7651	1.0464
Shanghai	6	6	1.4031	1.5404
Shanxi	30	21	–0.8516	–0.4185
Sichuan	12	12	–0.1431	–0.1431
Taiwan	1	1	2.2121	2.2121
Tianjin	24	13	–0.5133	–0.1619
Tibet	34	10	–2.0110	0.1422
Xinjiang	15	10	–0.2358	0.1195
Yunnan	17	16	–0.3252	–0.2531
Zhejiang	7	6	1.3717	1.6259

Source: Asia Competitiveness Institute

respectively after simulation. It is worth noting that the rankings of these economies tend to be low before simulation. On the other hand, Guangdong, Liaoning, Yunnan and Zhejiang have shown marginal improvement of one spot each after simulation. The Government and Institutional Setting ranking for nine economies have remained constant after simulation. They are Beijing, Fujian, Hebei, Jiangsu, Jiangxi, Shandong, Shanghai, Sichuan and Taiwan.

Although the rankings of Fujian, Jiangsu, Shandong and Shanghai have remained unchanged after simulation, their scores have improved. The reason for these four economies' inability to improve their respective rankings after the simulation lies in the large gaps in score between these economies and the economies ahead of them. For example, while the Government and

Institutional Setting score of Fujian has improved from –0.0487 to 0.2197 after the simulation, Fujian (#10) still lags behind that of #9-ranked Shandong in terms of score.

As for Beijing, Hebei, Jiangxi, Sichuan and Taiwan, both their scores and rankings are unaffected by the simulation, with their scores and rankings remaining the same. This is due to the absence of indicators of this environment within the top 20% weakest indicators of Beijing, Hebei, Jiangxi, Sichuan and Taiwan's.

The What-if Competitiveness Simulation Analysis on Financial, Businesses and Manpower Conditions is shown in Table 2.14. Economies that have shown major potential for improvements include Heilongjiang and Ningxia (each rises nine spots) and Anhui, Jiangxi, Sichuan, Tibet and Xinjiang (rises seven spots). On the other hand, economies that have improved by only one spot after the simulation include Hebei, Shandong, Shanghai and Taiwan. All economies in

Table 2.14: ACI 2015 What-if Competitiveness Simulation Analysis on Financial, Businesses and Manpower Conditions Ranking and Score

	Rank		Score	
Economy	**Before**	**After**	**Before**	**After**
Anhui	21	14	–0.4474	–0.1416
Beijing	5	5	1.1804	1.2883
Chongqing	15	15	–0.2166	–0.2166
Fujian	12	12	–0.0909	0.0919
Gansu	33	27	–1.0842	–0.8031
Guangdong	4	4	1.3441	1.4270
Guangxi	28	25	–0.8160	–0.5734
Guizhou	31	27	–1.0378	–0.7215
Hainan	27	22	–0.7900	–0.4631
Hebei	13	12	–0.1270	0.0962
Heilongjiang	22	13	–0.4740	–0.1283
Henan	17	12	–0.3237	0.0182
Hong Kong	1	1	2.8250	2.8250
Hubei	18	13	–0.3434	–0.1179
Hunan	20	14	–0.3925	–0.1514
Inner Mongolia	14	12	–0.2059	0.0738
Jiangsu	6	6	1.0780	1.1428
Jiangxi	26	19	–0.6430	–0.3549
Jilin	16	12	–0.2179	0.0313
Liaoning	10	10	0.2315	0.3718
Macau	3	3	1.4588	1.6563
Ningxia	30	21	–0.8820	–0.4508

(*Continued*)

Table 2.14: (*Continued*)

Economy	Rank		Score	
	Before	After	Before	After
Qinghai	32	27	−1.0613	−0.6726
Shaanxi	23	17	−0.5122	−0.2906
Shandong	9	8	0.4820	0.7939
Shanghai	8	7	0.7162	0.9338
Shanxi	25	21	−0.6029	−0.4302
Sichuan	19	12	−0.3583	0.0507
Taiwan	2	1	2.8019	2.8730
Tianjin	11	11	0.1581	0.2197
Tibet	34	27	−1.1726	−0.7242
Xinjiang	24	17	−0.5199	−0.3095
Yunnan	29	25	−0.8813	−0.5743
Zhejiang	7	7	0.9249	1.0175

Source: Asia Competitiveness Institute

the Greater China have improved their standing in this environment after simulation except for 10 economies, namely Beijing, Chongqing, Fujian, Guangdong, Hong Kong, Jiangsu, Liaoning, Macau, Tianjin and Zhejiang, which have retained their respective rankings even after simulation.

For the case of Beijing, Fujian, Guangdong, Jiangsu, Liaoning, Macau, Tianjin and Zhejiang, their lack of improvement in rankings can be explained by the large gaps in scores between their economy and the economy ranked above them. Although their scores have improved after simulation, the increase in score is still insufficient for them to surpass the score of the economy ranked ahead of them.

As for Chongqing and Hong Kong, their scores remain the same at −0.2166 and 2.8250 respectively even after simulation. This is because none of the indicators from the Financial, Businesses and Manpower Conditions environment is identified as the top 20% weakest indicators for Chongqing or Hong Kong. Furthermore, even if Hong Kong's score improved upon simulation, its ranking would not change as it already assumes the top position for the ranking of this environment.

Table 2.15 illustrates the 2015 What-if Competitiveness Simulation Analysis on Quality of Life and Infrastructure Development. For this environment, several economies have shown a promising potential for improvement. After simulation, Hong Kong rises 16 places; Henan rises 15 places; Sichuan and Yunnan each rises 14 places; Jiangxi rises 13 places; and Anhui and Guangxi each rise 12 places. With the exception of Beijing, Guangdong and Shandong, all the economies within Greater China have seen improvements in their rankings, including Hainan, Hebei, Jiangsu and Shanghai which have each improved one spot after simulation. Although

Table 2.15: ACI 2015 What-if Competitiveness Simulation Analysis on Quality of Life and Infrastructure Development Ranking and Score

	Rank		Score	
Economy	**Before**	**After**	**Before**	**After**
Anhui	23	11	–0.3881	0.4034
Beijing	1	1	2.3659	2.9861
Chongqing	31	20	–1.2171	–0.1980
Fujian	10	6	0.4571	0.8724
Gansu	32	26	–1.3518	–0.8116
Guangdong	2	2	1.6918	2.1908
Guangxi	27	15	–0.9368	0.1605
Guizhou	33	23	–1.5234	–0.3315
Hainan	26	24	–0.8991	–0.4354
Hebei	8	6	0.6600	1.1325
Heilongjiang	17	10	0.0120	0.5453
Henan	21	6	–0.1942	0.8670
Hong Kong	22	6	–0.2098	1.2874
Hubei	18	10	0.0005	0.5007
Hunan	24	13	–0.5739	0.3117
Inner Mongolia	12	6	0.3469	0.8931
Jiangsu	4	2	1.6156	2.2190
Jiangxi	30	17	–1.1050	0.1163
Jilin	19	15	–0.1190	0.1628
Liaoning	11	6	0.3935	0.9717
Macau	14	4	0.2208	1.5127
Ningxia	29	26	–1.1034	–0.6076
Qinghai	28	24	–1.0626	–0.5008
Shaanxi	16	11	0.1557	0.4066
Shandong	6	6	0.7669	1.3108
Shanghai	3	1	1.6167	2.3865
Shanxi	15	10	0.1691	0.4765
Sichuan	20	6	–0.1651	0.8641
Taiwan	7	1	0.7472	2.2254
Tianjin	9	5	0.5720	1.4032
Tibet	34	23	–2.0100	–0.4592
Xinjiang	13	6	0.2779	0.7756
Yunnan	25	11	–0.5814	0.4090
Zhejiang	5	2	1.3711	1.8451

Source: Asia Competitiveness Institute

Beijing is currently at the top for the Quality of Life and Infrastructure Development ranking, simulation analysis suggests the possibility of Beijing further enhancing its competitiveness in this environment. Beijing has the potential to improve its score from 2.3659 to 2.9861.

It is worth noting that rankings for Shanghai, Jiangsu and Zhejiang are #3, #4, and #5, respectively before simulation, but can rise to #1 for Shanghai and #2 for Jiangsu and Zhejiang after simulation analysis. This shows that Shanghai, Guangdong and Jiangsu are not far from #1 and #2-ranked Beijing and Guangdong or from each other in terms of Quality of Life and Infrastructure Development. Competition to obtain the top two positions is also very intense. The analysis shows that Shanghai, Jiangsu and Taiwan all have potentials to be ranked within the top two positions after the simulation. The key to emerge in the competitiveness race lies in which economy is able to improve their top 20% weakest indicators the fastest.

2.3 CONCLUDING REMARKS AND POLICY IMPLICATIONS

In summary, the top three performers for the ACI 2015 Overall Competitiveness ranking are Guangdong, Jiangsu and Taiwan, with the gaps in scores between the top ranked economies steadily diminishing over the years. The competition to remain at the top within the Greater China has intensified. Since 2010, economies from mainland China such as Guangdong and Jiangsu have surpassed Taiwan and Hong Kong in terms of Overall Competitiveness.

The relative strengths of the top three economies are shown in Figure 2.7, wherein the sub-environment scores of Guangdong, Jiangsu and Taiwan are compared to the maximum score of the 34 Greater China economies. For the #1-ranked Guangdong, it has almost achieved the maximum score of the 34 Greater China economies in the sub-environment of Government Policies and Fiscal Sustainability. Guangdong has also obtained scores close to the maximum in the sub-environments of Financial Deepening and Business Efficiency as well as Physical Infrastructure. As for the #2-ranked Jiangsu, it has achieved the maximum score of the 34 Greater China economies in the sub-environment of Regional Economic Vibrancy. Meanwhile, the #3-ranked Taiwan has achieved the maximum score of the 34 Greater China economies in the sub-environments of Labour Market Flexibility and Productivity Performance. However, the strength of Taiwan is rather unevenly distributed among the sub-environments. For example, it performs rather poorly compared to the maximum score in the sub-environments of Attractiveness to Foreign Investors and Physical Infrastructure.

The potential of Jiangsu, Taiwan and Hong Kong (currently at #2, #3 and #5, respectively in the 2015 Overall Competitiveness ranking) to replace Guangdong as the most competitive economy in Greater China is demonstrated in the What-if Competitiveness Simulation Analysis. It shows that these three economies can rise to #1 after the simulation analysis joining Guangdong. This shows that Jiangsu, Taiwan and Hong Kong are not far from #1-ranked Guangdong or from each other in terms of Overall Competitiveness. Top performing economies should not be complacent about their Overall Competitiveness standings. When each economy improves on their relative weaknesses while capitalising on their strengths, the competitiveness landscape within and out of Greater China will change. The key to emerge in this competitiveness race is the speed in improving the economy's weaknesses.

Figure 2.7: ACI 2015 Maximum Competitiveness Web — Guangdong, Jiangsu and Taiwan[24]

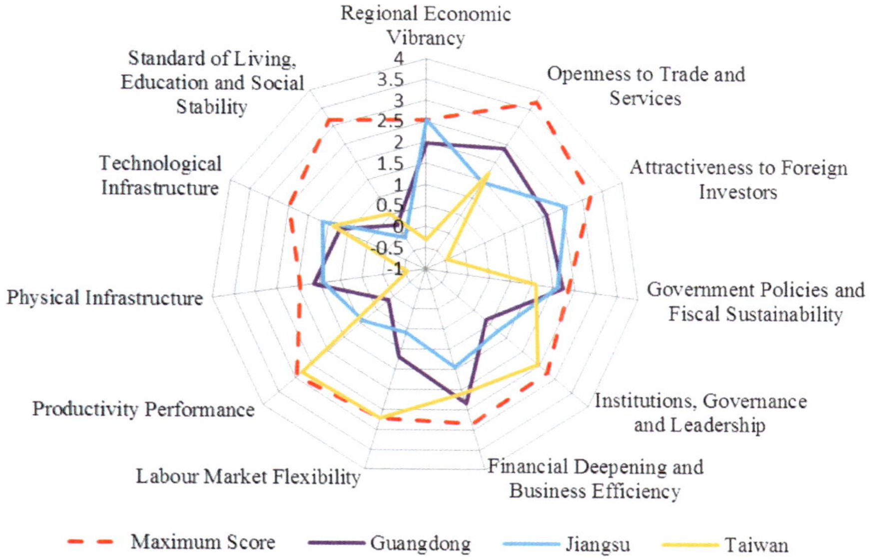

Source: Asia Competitiveness Institute

The bottom three performers of the ACI 2015 Overall Competitiveness are Guizhou, Gansu and Tibet. The rankings of these three economies have been fluctuating between #31 and #34 for the past 13 years (2000–2012), with their scores becoming increasingly negative, an indication of deteriorating competitiveness. Figure 2.9 is thus devised to further illustrate the relative strengths and weaknesses of these economies, whereby the scores of Guizhou, Gansu and Tibet in each sub-environment are compared to the median scores of the 34 Greater China economies.[24]

With the exception of Tibet which scores slightly above the median in the sub-environment of Openness to Trade and Services, all scores of the three economies are below the median scores. However, several sub-environments scores are close to the median scores, including Openness to Trade and Services and Attractiveness to Foreign Investors. Another observation of Figure 2.8 is that with the exception of Tibet which trails behind in the sub-environments of Institutions, Governance and Leadership as well as Physical Infrastructure, the scores of these three economies are largely close to each other. This implies that there is a healthy level of competition among the three economies in the majority of the sub-environments.

All these results paint a rather relatively bleak picture of the competitiveness of Guizhou, Gansu and Tibet as compared to the rest of the Greater China economies. However, based on the What-if Competitiveness Simulation Analysis on Overall Competitiveness, Guizhou, Gansu and Tibet have the potential to attain #25, #27 and #22, respectively should they choose to improve upon their top 20% weakest indicators to the average. This is indeed a very encouraging result. While the thrust of this ACI study has been focused on discovering the relative competitiveness

[24]Results shown in Figures 2.7 and 2.8 are based on 2012 data, collated and analysed in 2015.

Figure 2.8: ACI 2015 Median Competitiveness Web — Guizhou, Gansu and Tibet

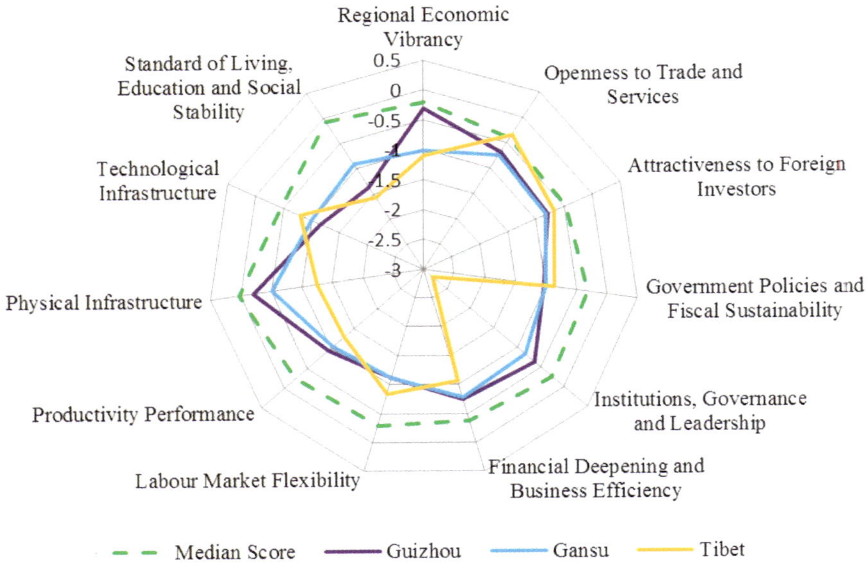

Source: Asia Competitiveness Institute

of the 34 Greater China economies, such policy-induced simulations offer useful insights for each economy to enhance its competitiveness, enabling the proposal of some useful development strategies or policies.

REFERENCES

Becker, G.S. 2007. "Health as Human Capital: Synthesis and Extensions." *Oxford Economic Papers* 59(3): 379–410.

Becker, Gary S. 1994. *Human Capital: A Theoretical and Empirical Analysis, with Special Reference to Education*. Third Edition. Chicago: University of Chicago Press.

Dollar, David. 2015. "China's Rise as a Regional and Global Power: The AIIB and the 'One Belt, One Road.'" *Horizons by The Brookings Institution* Summer (4): 162–172.

Friedman, Willa, Michael Kremer, Edward Miguel and Rebecca Thornton. 2011. *Education as Liberation?* Cambridge, MA: Cambridge University Press.

Li, Yining, Justin Yifu Lin, Yongnian Zheng, Yansheng Zhang, Min Tang, Monan Zhang, Zhicheng Bao, Jincheng Xiao, Xianjie Shen and Zhiyuan Sun. 2015. *Understand One Belt, One Road. Understand One Belt, One Road*. Beijing: CITIC Publishing Group.

Li, Zhinan. 2015. "Li Keqiang: Capitalise on the Upgraded Domestic Consumer Demands, Thereby Promoting Industrial Upgrading in China." *The Website of the State Council of the People's Republic of China*.

Liu, Yuanli. 2004. "China's Public Health-Care System: Facing the Challenges." *Bulletin of the World Health Organization* 82(7): 532–538.

MyCOS. 2015. "Self-Employment or Startup Continues to Rise: A Preview for the Trend of 'a Sinking Centre of Gravity.'" *Guangming Daily*.

Nataraj, Geethanjali and Richa Sekhani. 2015. "China's One Belt One Road: An Indian Perspective." *Economic and Political Weekly* 50(49): 67–71.

National Bureau of Statistics of People's Republic of China. n.d. "统计分类标准 Statistical Classification Standard (in Chinese)."

Perlez, Jane. 2015. "Answering an Appeal by Mao Led Tu Youyou, a Chinese Scientist, to a Nobel Prize." *The New York Times,* October 6.

Tan, Khee Giap, Randong Yuan, Wei Cher Sangiita Yoong and Mu Yang. 2014. *Annual Analysis of Competitiveness, Simulation Studies and Development Perspective for 34 Greater China Economies*. Singapore: World Scientific Publishing Company.

Tan, Khee Giap, Randong Yuan and Wei Cher Sangiita Yoong. 2015. *2014 Annual Provincial and Regional Competitiveness Analysis for Greater China Economies: Development Strategies under the New Normal*. Singapore: World Scientific Publishing Company.

The Economist. 2013. "Luxury Goods in China: Beyond Bling." *The Economist*, June 8.

Yu, Mu. 2015. "Premier Asks How Government Services Can Help to Meet the Upgraded Consumption Demands of the People." *People's Daily Online*.

Zhu, Xiaohui. 2015. "The Liberation of Women in the Initial Phase since the Founding of the People's Republic of China: A Human Rights Perspective." *Human Rights* 3: 89–100.

Discussant Notes 1

Thought on Promoting and Upgrading the Trade and Economic Cooperation between Shandong and Singapore in the Context of One Belt, One Road

Professor Zhang Shucun

Dean, Shandong Academy of Social Sciences
People's Republic of China

This year is the 50th anniversary of the founding of Singapore. In the past 50 years, Singapore sizes up the situation and constantly adjusts to changing international and domestic environments. It has transformed from a crowded and backward place with urban slums everywhere into a highly modern country. Although being small in terms of physical size, the scale of its achievements and the pace of its development have been tremendous. Shandong attaches great importance to developing trade and economic cooperation with Singapore. Bilateral trade and investment levels have been increasing steadily, with financial cooperation being the frequent highlights and new achievements constantly made arising from the collaboration in areas such as science and technology, education, environmental protection, human resources and social governance. A maturing institutional mechanism for the cooperation and remarkable achievements have resulted from such cooperation. But in recent years, the world economy has entered into a period of deep transition and adjustment, and China has put forward new developments and ideas in foreign economic cooperation. Under the new situation, the limitations of the original development path and model of the economic and trade cooperation between Shandong and Singapore have become increasingly prominent. The know-how to "start new perspectives, explore new areas, and discover new models" has become an urgent issue we face in the current economic and trade cooperation between Shandong and Singapore.

1. THE CURRENT STATUS OF THE ECONOMIC AND TRADE COOPERATION BETWEEN SHANDONG AND SINGAPORE

In the past 25 years, the economic and trade cooperation between Shandong and Singapore has progressed swiftly with a deepening friendly relationship and increased personal exchanges between China and Singapore under the backdrop of rapid regional economic growth and the further development in China–Association of Southeast Asian Nations (ASEAN) Free Trade Agreement.

1.1 The Rapid Growth of Import and Export Trade

In 1998, the total amount of the import and export trade between Shandong and Singapore was only US$ 173 million, of which exports were US$ 124.74 million and imports US$ 48.51 million. After a short period of 16 years, as of 2014, the total amount of imports and exports between Shandong and Singapore rose to US$ 3.28 billion, an increase of 18 times, including Shandong's exports to Singapore of US$ 1.98 billion, an increase of 14.9 times, and imports of 1.3 billion, an increase of 25.8 times. Such growth rates are much higher than the overall pace of the opening up of Shandong to foreign trade in the same period.

Despite the rapid growth, the trade volume between Shandong and Singapore is still small. As can be seen from Table 2.16, regardless of before or after the financial crisis, the share of the amount of imports and exports between Shandong and Singapore in China's trade with Singapore has been consistently below 6%, which is not congruent with the huge economic and trade development potentials in Shandong and Singapore, nor with Shandong's position in China's opening up process. It indicates an urgent need to break the bottleneck in the trade and economic cooperation between Shandong and Singapore in the future.

Figure 2.9: Trends in Total Imports and Exports between Shandong and Singapore Trade, 1998–2014, Unit: 10,000 US$

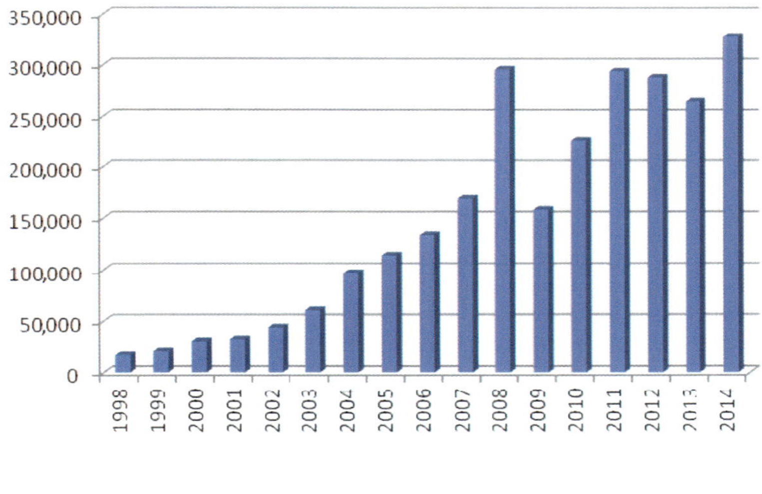

Table 2.16: The Share of the Trade between Shandong and Singapore in China's Trade with Singapore, Unit: US$ 10,000, %

Year	Total Amount of Imports and Exports Between Shandong and Singapore	As a Percentage in Total Amount of Imports and Exports Between China and Singapore	Total Exports from Shandong to Singapore	As a Percentage in Total Exports from China to Singapore	Total Imports from Singapore to Shandong	As a Percentage in Total Imports from Singapore to China
2007	169,130	3.59	82,773	2.79	86,356	4.93
2008	295,689	5.63	135,761	4.20	159,929	7.93
2009	158,179	3.31	63,635	2.12	94,544	5.31
2010	226,430	3.97	120,006	3.71	106,424	4.30
2011	293,783	4.61	176,820	4.97	116,963	4.16
2012	287,891	4.16	176,966	4.34	110,925	3.89
2013	264,641	3.49	168,201	3.67	96,440	3.21
2014	327,792	4.11	197,703	4.05	130,089	4.22
From January to July, 2015	157,000	3.36	99,000	3.29	58,000	3.49

Note: The data for January to July 2015 are estimates based on research

1.2 Playing an Important Role in China's Investments to and Cooperation with Singapore

In terms of two-way investments, the cooperation between Shandong and Singapore has achieved better results. In the past seven years, Shandong's actual use of funds from Singapore has accounted for one-tenth of China's actual use of funds from Singapore. In 2014, Shandong absorbed US$ 1.21 billion of investment from Singapore, an annual increase of 110% and accounting for one-fifth of Singapore's actual total investments in China. By the end of 2014, Shandong has accumulatively absorbed a total amount of US$ 7.34 billion actual funds from Singapore.

Shandong's investments to Singapore started relatively late, but rapid development has been achieved in recent years with very strong growth momentum. Shandong approved 19 companies in 2014 to invest in Singapore with US$ 200 million of approved overseas investments. As of the first half of 2015, Shandong approved 115 enterprises to invest in Singapore with the total approved foreign investments of US$ 1.41 billion.

1.3 Rapid Development in Labour Cooperation

Labour cooperation between Shandong and Singapore has made rapid development. Prior to 2005, Singapore was open only to us in construction sectors, but at the same time the companies in manufacturing, maritime and services industries were allowed to recruit a large number of

Table 2.17: The Actual Use of Funds from Singapore by China and Shandong, Unit: US$ 10,000, %

Year	Actual Use of Funds from Singapore by Shandong	Actual Use of Funds from Singapore by China	Ratio
2007	46,580	443,529	10.50
2008	46,667	318,457	14.65
2009	45,157	360,484	12.53
2010	54,996	542,820	10.13
2011	51,894	609,681	8.51
2012	70,947	630,508	11.25
2013	57,553	722,872	7.96
2014	121,000	58,000	20.86

Figure 2.10: Trends in Shandong's Investments to Singapore, Unit: US$ 10,000

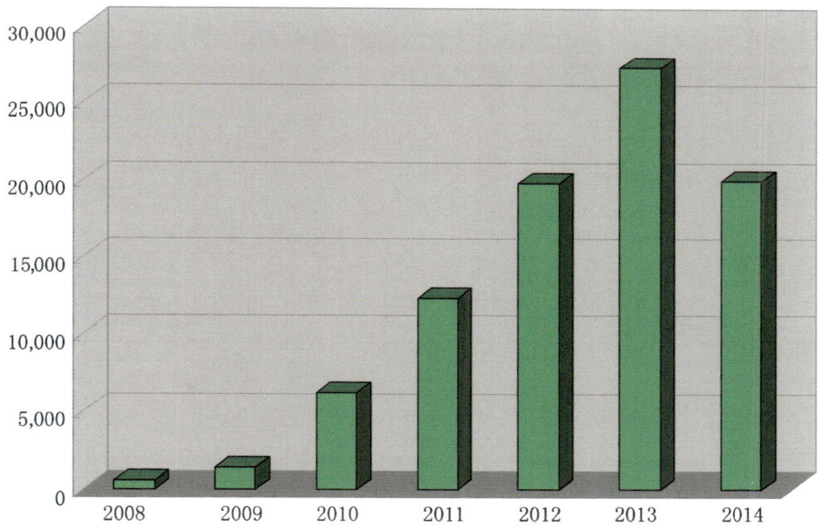

skilled industrial workers from China through special licensing arrangements. Since February 2005, Singapore opened the door to China for its labour market for manufacturing sectors, allowing its companies in manufacturing sectors (excluding the maritime industry) to hire workers from China up to the limit of 10% of the total number of their employees. Such imported labours were mainly concentrated in the construction, manufacturing, maritime and fisheries industries. The value of new contracts of foreign contracted projects between Shandong and Singapore in 2014 amounted to US$ 620 million, with a completed turnover of the foreign contracted projects totalling US$ 1.16 billion, with 6117 of all kinds of labours being sent from Shandong to Singapore in that year and 13,000 workers from Shandong staying over in Singapore by the end of the year.

After 20 years of development, the mechanism for economic and trade cooperation between Shandong and Singapore is maturing with increasingly close exchanges of people and remarkable achievements being made in such cooperation. Singapore has become Shandong's second largest overseas labour market, its third largest overseas project contracting market, its fourth largest source of foreign investments and its ninth largest overseas investment destination.

2. THE CHARACTERISTICS OF THE COOPERATION BETWEEN SHANDONG AND SINGAPORE

Because of the special conditions in the actual process of our economic development, the "industrial park model" as represented by the Suzhou Industrial Park did not take place in the process of cooperation between Shandong and Singapore. Instead, under the guiding idea to make full use of Singapore's resources and to actively draw on Singapore's experiences, we pursued a path of "diversified development out of the industrial park". Such a development model is illustrated mainly in the following three aspects.

2.1 Attach Great Importance to the Economic Zones, but not Confined to the Economic Zones

Singapore has a wealth of experience in the planning and construction of the industrial parks. Strengthening economic cooperation in building and managing economic zones and industrial parks has always been one of the major areas of the economic and trade cooperation between Shandong and Singapore. It was proposed during the Singapore–Shandong Economic and Trade Council held in 2014 that "the two sides will make joint efforts in full support of accelerating the process of cooperation in the China–Singapore Jinan Knowledge Industry Park, high standard theme park planning, industry position, leading the industry work together to build a new generation of industrial restructuring and development Park". Furthermore, we should introduce to Shandong from Singapore the advanced planning methods for industrial parks, philosophy in attracting investments and development models, in order for Shandong to boost its efforts in transforming and upgrading its economic zones and industrial parks. But beyond the cooperation in industrial parks, Shandong and Singapore have also launched many other programmes of the exchanges and cooperation, mainly reflected in the active introduction of Singapore's experience to help Shandong in reforming its institutional mechanisms, building a service-oriented government, vigorously improving the business environment, creating new advantages for Shandong as an open economy.

2.2 Focus on International Trade and Economic Cooperation, but Stick to a Diversified Cooperation Strategy

Singapore is a developed economy with advanced technological capabilities and scientific ways of management, and Shandong has rich natural resources and labour resources. There is a high level of structural complementarity between the industries in Shandong and Singapore and a huge potential for the international economic and trade cooperation, which will provide an incessant power to promote the development of the bilateral cooperation. The fact that there has been a rapid development in recent years in the economic and trade cooperation between Shandong and Singapore clearly illustrates such situation. As Singapore is an advanced industrialised country, it has in the process of its past development not only made great economic achievements, but also accumulated the valuable experiences in the development of its society, in areas such as urbanisation, social governance, civil servants management, state-owned enterprises reform and so on. In the course of their past cooperation, Shandong and Singapore have collaborated widely in many of the above-mentioned areas while conducting their economic cooperation.

2.3 Actively Undertake Industrial Transfer, but Pay Special Attention to the Quality of the Economic Cooperation

Since the 1980s, the government of China has further defined the development strategy for China's reform and opening up, to attract foreign direct investment and develop the processing trade. China's manufacturing industry has therefore entered a period of rapid growth. Relying on the advantage of cheap labour, land and other primary factors of production, China's manufacturing industry has joined and participated deeply in the division of labour in the global value chain, and gradually grown into the world's manufacturing base. Under the impact of the general trend, the economic cooperation between Shandong and Singapore also carries the above characteristics to a certain extent, reflected from the fact that the multinational companies in Singapore will transfer the labour intensive processes to Shandong at cheaper prices but will keep the knowledge-intensive processes in Singapore. But in recent years, the quality of bilateral cooperation has been significantly improved. A number of Singapore-based multinational companies began to set up regional headquarters in Shandong. The pace at which large projects are brought into Shandong has significantly accelerated. There has been a trend of accelerated development in the cooperation in producer services, especially in research and design, information technology services, energy saving and environmental protection, business consulting, outsourcing services, brand building and etc.

3. THE NEW SPACE FOR THE TRADE AND ECONOMIC COOPERATION BETWEEN SHANDONG AND SINGAPORE UNDER THE NEW SITUATION

In 2013, during his visit to Central Asia and Southeast Asia, Chinese President Xi Jinping, proposed respectively two initiatives in building the Silk Road Economic Belt and the 21st Century Maritime Silk Road, imparting a brand new connotation of the times to the ancient Silk Road. Promoting the construction of the Silk Road Economic Belt and the Maritime Silk Road is a strategic development to deal with the current environment at home and abroad, which embodies China's new philosophy of promoting world economic development with a thrust in openness, freedom and cooperation. It is also an important development strategy for China to promote inclusive regional growth and to build a new pattern of an all-round opening up of the country, which is the major initiative for China to deepen its reform and opening up in particular towards the west and the south. Both the land and sea routes of the new Silk Roads comply with the new changes in the economic, political and diplomatic structure of the world. It will also further promote exchanges and cooperation between China and neighbouring countries, in order to stimulate the vitality of development for each country, to create mutually beneficial relationships and win–win situations in broader fields, and to create new opportunities for long-term stability and prosperity in the region. This will also provide a new space for the partnership between Shandong and Singapore.

3.1 Singapore Has a Long and Valuable Experience in Exploring and Developing Its International Markets

Compared with Western countries, Singapore was founded with a relatively short history, and the course of its outward investment in foreign countries is also not very long. But it has achieved good results. During the economic take-off of Singapore, outward investment has not only eased the problem of small markets and the shortage of resources it faces in its own country, but also brought sizeable national incomes into Singapore. Singapore has invested extensively overseas, collaborating with many countries. In addition to China, countries in ASEAN, Europe, Central and South America are also its important partners. The cooperation with ASEAN is mainly focused on the integration of industrial chains. The cooperation with Europe is mainly market development-oriented. The cooperation with Central and South America is mainly resource development-based. Its selection for partners and fields of cooperation in its overseas investment is well matched with the development path of Shandong's outward investment under the "One Belt, One Road" strategy. Strengthening the cooperation with Singapore, and bringing together the outward investment and cooperation, based on the advantage in Shandong's production capacity and the experience of Singapore in outward investment and cooperation will produce positive effects for the both sides.

3.2 Singapore Has Valuable Experience in Building the Economic Zones and Industrial Parks in Developing Countries

In the process of Singapore's outward investment and cooperation with foreign countries, the construction of economic zones overseas has always been an important aspect of its success. As early as in the early 2000s, the Singapore government proposed the "Regionalization 2000" programme, indicating the cooperation in industrial parks as an important part of its economic development. In a period of 20 years, Singapore established many overseas industrial parks in Indonesia, Vietnam, China, India and other countries. In the process of development of these parks, Singapore has always been adhering to the same principles in construction design and organisation. Singapore has actively introduced its standards in quality infrastructure and its philosophy of clean and efficient administrative management. This has helped create a business environment in the host country which is similar to Singapore's own business environment that is conducive for foreign investors, thereby enhancing the ability of the host country to attract foreign investment. As a result, Singapore has been warmly received by host countries and investors. Shandong and Singapore not only can strengthen their cooperation in the development of the economic zones in Shandong, but also can strengthen exchanges and collaboration in building industrial parks in third countries, in order to create a first-class environment and competitive economic zones.

3.3 Singapore Has a Good Investment Environment

Singapore may in the future become the "springboard" for enterprises in Shandong to go abroad. Singapore has trading partners around the world, a more developed logistics system, and a very important strategic location. Singapore has signed 21 Free Trade Agreements with 32 trading partners and signed 41 investment guarantee agreements and 76 agreements to avoid double taxation with its key partners. There are more than 600 local and overseas financial institutions operating in Singapore, where more than 800 law firms and a large number of accounting firms are doing business, which can provide good support for the enterprises going abroad. Enterprises from Shandong can leverage on Singapore with an open and excellent investment environment, via Singapore sailing abroad. They can use Singapore as an international trade platform to enjoy Singapore's open and free trade environment, excellent location, lower financing costs and taxes. They can also set regional headquarters in Singapore, enjoying the tax advantages and the fair treatment in investment. Singapore also possesses an effective capital market, and companies from Shandong can optimise Singapore's capital market for their financing needs in order to open up the markets along the "One Belt, One Road". For example, if Singapore signs the investment agreement with the host country, we can avoid double taxation and enhance investment protection, set up specialised agencies to provide the companies with a range of services, to provide tax, credit and insurance support for their overseas investment. We can also encourage enterprises to hold together to go abroad and speed up the aggregated development in their outward investment.

4. DEVELOP A NEW SPACE AND CALL FOR A NEW MODEL OF COOPERATION

The new space for cooperation has brought new opportunities for the economic cooperation between Shandong and Singapore. But the original model of cooperation has a lot of "inward looking" features, mainly reflected by the established institutional mechanism for Shandong to learn from Singapore's valuable experience to promote the construction of economic zones and the formation of an industrial chain boosted by absorbing investment from Singapore to Shandong. The "One Belt, One Road" initiative has been supported by the majority of the relevant countries. Therefore, the countries and regions along the "One Belt, One Road" will become a rapidly rising industrial belt for international trade and economic cooperation. Under the new situation, Shandong and Singapore should hold a long-term view, to strengthen cooperation in broader areas, to bring forth more impetus for the development and benefits for both sides. Therefore, under the new situation, the new model of cooperation between Shandong and Singapore should not only include the existing cooperation in industrial parks and economic zones and the considerations to promote institutional innovation inside the province, but should also focus on jointly exploring the international market. To jointly create a community of countries along the "One Belt, One Road" which share common interests, should be the new content in the model of cooperation between Shandong and Singapore in the future. Compared with the previous model of cooperation, this new paradigm should include the following four aspects.

4.1 Transition from Drawing on the Experience from Singapore to Construct the Economic Zones in the Province to Jointly Building Industrial Parks Overseas

Planning and building the industrial parks is where the advantage lies for Singapore. The proposal of the One Belt, One Road initiative has not only brought about opportunities to develop overseas industrial parks but also demanded acceleration in development. The accelerated construction of the industrial parks overseas is required for our province to operate with economies of scale in the industries of our outward investment. It is also needed to mitigate the risks associated with the outward investment. Shandong and Singapore should strengthen the cooperation in building the industrial parks in foreign countries, in order to enhance the salient features and the operational execution power in Shandong's industrial parks overseas and to further improve the environmental competitiveness of the overseas industrial parks.

4.2 Transition from Absorbing Funds from Singapore to Complement the Industrial Chains in Our Province to Promoting the Relocation of Industrial Chains into Foreign Countries

In the past economic cooperation, Shandong attracted a large number of investment projects from Singapore. These projects have helped to elongate the industrial chains in Shandong to a certain extent to achieve the integration of industrial chains. In September 2013, Chinese President Xi Jinping pointed out in his speech in Kazakhstan that the goal of the initiative to jointly build the "Silk Road Economic Belt" is to promote the integration among the countries along the belt and establish a development space with closer economic ties. This means capitalising on the advantages in political relationship, geographical closeness and economic complementarity between China and the surrounding countries and converting those advantages into pragmatic cooperation and sustained growth, thereby creating a win-win situation for a community of countries with shared interests and a common destiny. Based on this, exploring new modes of industrial cooperation has become an important task in building the "One Belt, One Road". Shandong should conform to the latest trends in international economic development in the process of building the "One Belt, One Road", combine the use of foreign capital and its own outward investment, and jointly promote the extension of the production chains from Shandong to the countries along the "One Belt, One Road" through an overall planning. Singapore also has advantages in promoting the outward extension of the industrial chains.

4.3 Transition from Improving the Internal Business Environment to Building a Competitive Advantage in "Going Abroad"

Singapore is well known for its excellent business environment. In the annual Ease of Doing Business Report by the World Bank, Singapore was ranked at the top for business environment for nine consecutive years from 2006 to 2014. Singapore has long been adopting a set of business-friendly policies, "attaching great importance to investor services, establishing 'one-stop' service centres, streamlining the process for approval of projects, and minimising administrative fees", which has been recognised as a valuable experience and a role model by Shandong. In our cooperation with Singapore, Shandong continues to take measures to optimise the environment for economic development. Such measures include: in terms of utilising foreign capital, constantly improving the level of foreign investment facilitation, smoothing communication channels with foreign investment enterprises, and building an integrated service platform for foreign investors and so on; in terms of foreign trade, improving trade facilitation, providing innovative currency exchange management services, and increasing the efforts in international market development and so on; in terms of the transformation of government functions in management, continuing to regulate administrative matters, decentralising some items in administrative check and approval, and improving macroeconomic management and so on. Designing and implementing these measures has made a positive impact for optimising the business environment for Shandong, and enhanced the international competitiveness of the industries and products from Shandong. However, under the new initiative of the "One Belt, One Road", Shandong is facing the important task of strengthening the cooperation in outward investment. Only improving the province's internal business environment is not enough. How to build a competitive advantage to go abroad has become an important aspect of the strategy that has the impact on Shandong's integration into the "One Belt, One Road". In this regard, Singapore has already gained valuable experience, which brings a new space for development in the cooperation between Shandong and Singapore in the future.

4.4 Transition from Cultivating Economic Management Talents that Are in Shortage in the Province to Fostering Talents with International Experience

While Singapore is exporting its industries and capital, it is also actively exporting management experience. Training of talents has always been an important part of the economic and trade cooperation between Shandong and Singapore. In 2013, International Enterprise Singapore, together with seven educational institutions including National University of Singapore, jointly established the "Singapore Talent Development Alliance", providing one-stop education and training programmes for Chinese provinces and cities including Shandong and developing a full range of targeted education and training courses. The contents of the courses include all aspects related to development of the industrial parks, modern services industry, and government management and so on. The trainees include staff from government, senior management staff of businesses and even junior level employees. Training of the above personnel has to a certain extent, met the needs in the development process for China's industrialisation, informatisation and internationalisation, which has contributed to building a solid, long-term and sustainable partnership between mainland China and Singapore. As the economic globalisation continues to deepen, Shandong and the world economy are more closely linked, and Shandong experiences increasingly strong demand for international talents. We can say that talent has become the biggest problem in enterprises from Shandong to go abroad. Singapore has a hands-on experience in outward investment cooperation with a more advanced educational system. This complements well with Shandong, which is facing a reality of a shortage of talents. Training talents with international experience who can manoeuvre with ease in the international markets will become an important aspect of the partnership between Shandong and Singapore in the future.

5. MEASURES AND SUGGESTIONS TO PROMOTE THE TRANSFORMATION AND UPGRADING OF THE ECONOMIC AND TRADE COOPERATION BETWEEN SHANDONG AND SINGAPORE

5.1 Jointly Build a Platform for Industrial Cooperation Between Shandong and the Countries Along the "One Belt, One Road"

We recommend establishing a special committee for industrial cooperation in the Singapore–Shandong Business Council, to communicate and coordinate the relevant departments in the governments of Shandong and Singapore to promote industrial cooperation and joint development, in order to achieve the transformation and upgrading of the existing platform for industrial cooperation as soon as possible. We should seek further extensions for the functions of the industrial cooperation platform. The functions of the platform should not be limited to introducing to the foreign investors the status of economic development and investment opportunities in Shandong, helping foreign investors understand the markets in Shandong, and promoting the investment in Shandong by overseas investors including Singaporean investors. The functions of the platform should also include introducing to the entrepreneurs in Shandong the situation of the economic development in the countries along the "One Belt, One Road", the countries' measures in attracting investment and the investment opportunities in these countries, to help entrepreneurs in Shandong open up the overseas markets. When the timing is right, we should build a website for promoting investment in the countries along the "One Belt, One Road", and introduce international investment opportunities and conduct investor matchmaking based on the website.

5.2 Jointly Build a Platform for Training International Talents

We recommend that National University of Singapore and Shandong Academy of Social Sciences strengthen their cooperation, give full play to their respective advantages, jointly build a training centre for talents, and improve the talent training mechanism. In terms of the contents of the training programmes, we should follow and serve the needs for Shandong in strengthening its cooperation in investment with the countries along the "One Belt, One Road". On the one hand, we want to help shape the international strategic awareness of entrepreneurs from Shandong, establishing the awareness to seek active integration into the world and to participate in international competition. On the other hand, we want to help the entrepreneurs to form the awareness of corporate social responsibility, establishing an international image of "altruism" and "win–win" situations in the host country and eliminating the distrust of the host country in enterprises from Shandong.

5.3 Jointly Build a Platform for Trade Cooperation

We should hold international cooperation summits in which enterprises from Shandong participate to help companies from Shandong make their voices heard, establish a club for cooperation covering think tanks, research institutes and entrepreneurs from Shandong, advise the companies in their cross-border cooperation, and hold various types of trade fairs to introduce the opportunities for cooperation with the countries along the "One Belt, One Road". Furthermore, we should create a platform for providing trade information, which should include the information about the host country's business environment, such as the impact on imports and exports by the host country's national economic development planning and social development strategy, the trade opportunities brought about by new economic policies or plans, industrial upgrading planning and policy changes in industrial park management, and the information that may have implication for the profitability of the foreign trade activities by the enterprises from Shandong and so on.

5.4 Promote the Development of the Overseas Industrial Parks After Strengthening the Cooperation in Industrial Parks in the Province

Shandong and Singapore should give full play to their respective advantages, strengthen the cooperation, and jointly promote the development of overseas industrial parks and economic zones. On the one hand, Singapore should attach great importance to and actively participate in the construction of Shandong's overseas industrial parks using a variety of flexible methods. This can be either providing intellectual support for the development of Shandong's industrial parks in foreign countries, helping enterprises from Shandong in planning and designing the overseas industrial parks in the countries and regions along the "One Belt, One Road", or investing in foreign countries and building overseas industrial parks by companies from Singapore which capitalise on their advantages such as their knowledge in the local situation of economic development and their long history of cooperation with the countries and regions along the "One Belt, One Road", to attract companies from Shandong, especially those small to medium-sized companies, according to the needs of enterprises in Shandong. On the other hand, Singapore should provide supporting services for the development of Shandong's overseas industrial parks. Singapore's enterprises should actively participate in the development of ancillary services for Shandong's overseas industrial parks, providing professional planning and consulting services, financial and insurance services, as well as logistics services for enterprise in the parks, so that Shandong's overseas industrial parks can have a high starting point and a high construction standard to achieve healthy and sustainable development.

5.5 Guide the Extension of Industrial Chains to Singapore in the Cooperation of Key Industries

Centred on the key industries of cooperation between Shandong and Singapore, we should build cross-border industrial chains that are mutually beneficial to Shandong and Singapore. The first key industry is the producer services industry. Shandong and Singapore should take the producer services industry as a new starting point, to strengthen the investment cooperation in finance, insurance, information technology, modern logistics, services outsourcing, exhibition economy, e-commerce, consulting services, accounting and legal consulting, brand building, human resources management and other fields of the modern services industry. The second key area is the backbone competitive industries. Shandong and Singapore should actively promote the cooperation in novel pharmaceuticals, marine economy, medical equipment, high-performance fibres, specialty polymers, new materials, new metallic materials, functional ceramics and other fields. The third key area is major infrastructure construction. We should encourage Temasek and other companies from Singapore to participate in the construction projects of Shandong's highways, railways and other major infrastructure through public–private partnership and other models, support Singapore's enterprises to participate in the reform of state-owned enterprises in Shandong, and establish a diversified and sustainable mechanism for utilising foreign capital. We should encourage the joint development of major infrastructure by companies from Singapore and Shandong in the countries along the "One Belt, One Road". While promoting the cooperation in these areas, we should also further strengthen the policy coordination between Shandong and Singapore, draft mutually consistent support policies for the construction of industrial chains, enhance the enterprises' understanding and grasp in the industrial chains, and improve the enterprises' mastery in the design, management and control of the industrial chains and value chains.

5.6 Help and Support the Development of Shandong's Multinational Corporations

Firstly, we should encourage large enterprises to carry out the mixed ownership cooperation by introducing strategic investors from Singapore. We should give full play to Singapore's advantages in hosting many large enterprises and especially many of the Asia-Pacific regional headquarters for Fortune 500 companies, to attract the big companies and Singapore's sovereign wealth funds as well as other financial institutions to cooperate with state-owned enterprises in Shandong through mergers and acquisitions, cross-shareholdings and capital operations, to attract large companies from Singapore to set up regional headquarters, research and development centres and marketing centres in Shandong, and to enhance the quality and efficiency of Shandong in utilising Singapore's investment. The second is to encourage Shandong's enterprises to be listed in Singapore's stock exchanges or to issue bonds there, and to establish an outward investment base in Singapore. When the timing is right, we should establish in Singapore an aggregated area for regional headquarters for Shandong's enterprise. The third is to establish the mechanism for promoting business cooperation, to actively support the enterprises to establish research and development alliances, technology standard alliances, marketing cooperation alliances and industrial chain cooperation alliances, and to promote the integration of the development of enterprises in Shandong, Singapore and the countries along the "One Belt, One Road".

Discussant Notes 2

The "Thirteenth Five-Year Plan": Trends and Challenges in China's Economic Restructuring and Upgrading

Professor Chi Fulin
President, China Institute for Reform and Development &
Member of the Chinese People's Political Consultative Conference
People's Republic of China,
Delivered on Behalf of Professor Chi Fulin by Dr He Dongni
Vice President, China Institute for Reform and Development
People's Republic of China

During the Thirteenth Five-Year Plan, China will face a critical period for transforming and upgrading its economy. The hope to resolve the short-term growth pressures lies in 2020. The key for transforming the model of economic development lies in 2020. The critical point for achieving a comprehensively well-off society and joining the ranks of high-income countries lies in 2020. In this particular context, the Thirteenth Five-Year Plan should be based on the basic trends of economic transformation and upgrading, to form a good grasp of and a good plan for the selection of economic growth paths, to effectively respond to the challenges of economic restructuring, and to achieve substantive breakthroughs in economic transformation and upgrading.

1. THREE MAJOR TRENDS IN CHINA'S ECONOMIC TRANSFORMATION AND UPGRADING

1.1 The Major Trend in Industrial Restructuring and Upgrading: From "Made in China" to "Intelligent Manufacturing in China"

We face a historical intersection where the transformation in development model, the structural adjustment and the new round of global technological revolution meet each other. To grasp this important opportunity to initiate the transformation and upgrading of the structure of secondary industries is essential for the long-term economic development.

Secondary industries show prominent features of informatisation and being service-oriented. Secondary industries and information technology have become highly integrated, with secondary industries being increasingly individualised and service-oriented. The recently released programme of action "Made in China 2025" has reflected the major trend in shifting towards intelligent manufacturing, with the upgrading from Industry 2.0 to Industry 3.0 expected to be initially completed by 2020, which will lay the foundation for the transition towards Industry 4.0.

The "Internet Plus" has outstanding advantages for upgrading manufacturing industries. The ability to seize new opportunities in this new round of industrial revolution, to take full advantage of the initial formation of the business model innovation in "Internet Plus" and to promote the transformational change in manufacturing has a significant impact on the realisation of the strategic objectives for "Made in China 2025".

1.2 The Major Trend in Transformation and Upgrading of Urbanisation: From Scale Urbanisation to Population Urbanisation

China's population urbanisation rate will reach around 50% by 2020. In 2014, China's scale urbanisation rate reached 54%, and it is estimated that by 2020 it will reach 60%. The problem is that the population urbanisation rate is too low, only 36% in 2013. To reduce the gap between the population urbanisation rate and the scale urbanisation rate, we need to carry out greater reform in the household registration system, the equalisation of basic public services, the integration of urban and rural areas and other institutional innovations, in order to reach a population urbanisation rate of about 50% by the end of the "Thirteenth Five-Year Plan", which should be set as an important binding target for the new development model of urbanisation in the "Thirteenth Five-Year Plan", thereby releasing enormous hidden dividends through population urbanisation.

Let the term "migrant workers" become past history by 2020. This is not only an economic problem but also a major social problem. Generally speaking, 2020 is the critical point to make the unreasonable system of "migrant workers" take an exit from the stage of history. This is not a conditional problem, but rather an issue of determination.

It is plausible to achieve an annual increase of 2% in the population urbanisation rate during the "Thirteenth Five-Year Plan". First, with the acceleration of reform in the household registration system, the process of urban-rural integration will significantly accelerate. Second, the government of China has increased public resources available to the small and medium towns and cities, making them an important carrier to absorb the transfer of rural population. Third, the development of China's basic public services system is accelerating.

1.3 The Major Trend in Upgrading Consumption Patterns: From Material Consumption Towards Services Consumption

Consumers in China are upgrading from subsistence spending to development-oriented spending. A decade ago, consumer spending for subsistence accounted for a large proportion of total consumer spending for the urban and rural residents. In the past 10 years, the proportion of development-oriented consumption has increased significantly, such as education, healthcare, fitness and tourism and so on. For example, tourism used to be a niche, high-end market, but it has now become a common consumption item for the ordinary people.

Consumers in China are upgrading from traditional consumption goods to new types of consumption goods. In recent years, the demand for green consumption, information consumption, convenience consumption and other types of new consumption has further increased. Despite the intensifying downward pressure on the economy, the annual growth rate of consumption in communication equipment remained above 30% in the first half of 2015. In the past ten years, information consumption by urban and rural residents has increased at an annual rate of about 20%.

Consumers in China are upgrading from material consumption to service-oriented consumption. At present, the proportion of service consumption in the total consumption by urban residents is close to 40%, and it is expected to reach above 45% by 2020 and in some developed regions likely to reach around 60%.

The major trends in economic transformation and upgrading have shown that there is a tremendous potential in the next five to 10 years of economic growth in China. The pattern of an improving economic situation has not changed. The "China Story" has not finished. There is still much room for imagination.

2. THREE CHALLENGES FOR CHINA'S ECONOMIC TRANSFORMATION AND UPGRADING

2.1 Adjustment in Industrial Structure: From Being Led By Secondary Industries to Being Led By Services Industries

China's current economic structure is not compatible with the stage of its economic development. Currently, it has been clearly marked by indicators such as industrial structure, employment and employment contribution rate that China has entered the middle to later stages of industrialisation. By 2020, the services sector will account for more than 55% of the economic output, forming the basic pattern of an economy mainly led by services industry, which will be an important sign indicating the adjustment in the industrial structure during the middle to later stages of industrialisation. At present, although the amount of value added by services industry has increased significantly, its proportion in the GDP is still low, at only about 50% by the first half of this year. This is about 20% points lower than the developed countries, and about 10% points lower than countries such as India, Brazil and Russia.

Contradictions and problems are becoming more pronounced in the restructuring process. For example, labour costs are significantly increased, which is one of the reasons for the decline in the competitiveness of manufacturing industries. Constraints due to resources and environmental factors are making it impossible to sustain the types of manufacturing industries that are characterised by high consumption of energy and resources with relatively low efficiency.

To enhance China's international competitiveness requires forming an industrial structure that is mainly led by services sector as soon as possible. The proportion of the producer services industry that focuses on research and development is too low, which has become a major factor restricting the transformation and upgrading of the manufacturing sector. The "Thirteenth Five-Year Plan" should strive to increase the share of the producer services industry from 15% to 30% or even 40% of GDP.

2.2 Improving Innovation Capability: From Investment-Driven to Innovation-Driven

The key to being innovation-driven is the capability in scientific and technological innovation. Currently, China's capability in scientific and technological innovation is in the lower-middle level internationally. As shown in the survey conducted in 2013 by the National Leading Group for Reform in Scientific and Technological Research, in the 1,149 key technologies from the 10 sectors that participated in the survey, 195 of them (17%) reached the internationally advanced level, 355 of them (31%) reached the same or similar level as the internationally advanced level, and 599 of them (52%) were in the tracking phase with a level substantially lower than the internationally advanced level. In particular, the key technologies and core technologies had a foreign dependency ratio of as high as 60%.

China should rely on the "Internet Plus" to speed up the formation of the institutional environment for innovation and entrepreneurship. The "Internet Plus" is a historical opportunity for China to achieve technological catch-up. China has accumulated some major experiences in this area. For the next step, we need to speed up the formation of the institutional environment for the "wide participation of the people in innovation and entrepreneurship", releasing the enormous potential brought about by the "Internet Plus". For example, we can increase the size of the national development fund for small and medium-sized enterprises (SMEs), support and encourage more local governments to establish SMEs development funds. We should also guide local governments, venture capital institutions and other social funds to support SMEs and micro-enterprise development in their start-up period, and build an institutional platform for the innovation and entrepreneurship activities by the SMEs.

The foundation of the innovation-driven economy lies in education. The current education system in China has an overemphasis on examinations, and is a closed system with a heavy administration that lacks economic incentives, which in turn is an important constraint for forming independent innovation capability. If we cannot accelerate the formation of an innovative, open and professional education system, our independent innovation will lack the important basic conditions, which will make it difficult for us to meet the development needs for China to enhance its independent innovation capability. It will also be difficult to provide the various professionals required for the economic restructuring. Deepening the education reform during the "Thirteenth Five-Year Plan" requires us to "look at education from outside of education" and to accelerate the second education reform with the fundamental objectives of optimising the structure and improving the quality.

2.3 Comprehensive Breakthroughs in Reform: From Being Changed Passively to Seeking Active Changes

There must be an internal reform impulse and a conscious reform action. The Third and Fourth Plenary Sessions of the 18th Central Committee of the Communist Party of China emphasised the comprehensive reform. But the comprehensive reform is facing enormous challenges in interests adjustment and power structure adjustment, encountering great resistance for every item of the reform, which is much more profound, complex and arduous than the reform in the past 20 years. Today, the comprehensive deepening of the reform is very difficult, the key reason is that there are too many storytellers, too few people doing real work of reform, and fewer still who are dedicated to the reform with the drive and enthusiasm.

We should create the great atmosphere for reform. At this period when the needs for growth, restructuring and reform meet each other, to make the reform not going backward but accomplished successfully, we need to create the great environment for reform as soon as possible, to further emancipate the mind, to establish an incentive mechanism for reform, and to encourage reform pilot areas across the country. I am relatively optimistic in this regard. After visiting a lot of places, I found that many of them have practised their own "freestyle" reform actions in addition to the "routine" reform actions, some of which have become worthwhile examples to be promoted all across the country.

3. THREE MAJOR RELATIONSHIPS IN CHINA'S ECONOMIC TRANSFORMATION AND UPGRADING

3.1 Speed and Structure: From More Emphasis on Speed to a Greater Emphasis on Structure

Room in economic growth speed should be reserved for structural adjustments during the "Thirteenth Five-Year Plan". The economic transformation and upgrading are at a critical stage during the "Thirteenth Five-Year Plan". While maintaining an economic growth rate of 6%–7%, we should make structural adjustments as the focus of our economic reform. The increased pressure from the current economic downturn is due to both growth cycles and more importantly the problem in the development model. Objectively speaking, most of these contradictions and problems are directly related to the lagging economic restructuring. We should not overlook the objectives of changing the development model and restructuring the economy just because of the intensifying short-term downward pressure on the economy.

The key of the structural adjustments is to accelerate the development of modern services industry. The overall judgment is that we should shape the industrial structure such that the services sector plays a major role, which will induce and impact on the overall situation of the transformation and upgrading of the economy. First, in recent years, each one percentage point of growth in China's services sector has contributed to about 0.4% point of growth in GDP. If the services sector grows 10% annually during the "Thirteenth Five-Year Plan", it will contribute to 4% points of economic growth, laying an important foundation for the medium-speed growth. Second, each one percentage point of growth in the services sector currently can absorb about one million new jobs. If the services sector can grow at an average annual growth rate of 10%, it will absorb about 10 million new jobs every year in the next five years, which will be a main channel of attracting employment and open up a huge market space for entrepreneurship and innovation. Third, if by 2020 the services sector will account for 55% of GDP, energy consumption will fall by about 14%, and sulphur dioxide emissions will decrease by about 18%, which will effectively get rid of the path dependence in the heavy chemical industry, thereby creating favourable conditions for the realisation of the green development under the New Normal.

There exist the right conditions for creating an industrial structure mainly led by the services industry during the "Thirteenth Five-Year Plan". It is estimated that by 2020 the proportion of the services sector in GDP will reach more than 55%. Such estimation is based primarily on the following reasons. First, the services industry has entered a phase of rapid development, growing at an annual average rate of 10.6% from 2001 to 2013 and accounting for almost 50% of GDP by the first half of 2015. Second, there is a huge space for investment in the services sector, as demands for education, healthcare and other services are far from being met. It will require at least 10 trillion yuan of additional investment in order to achieve the overall equalisation of basic public services by 2020. Third, more than 80% of the secondary industry is in manufacturing, which is a highly market-oriented sector, while more than 50% of the services sector is still monopolised by administrative power. Once the market liberalisation accelerates in the services sector, huge growth potentials will be released.

3.2 Short-Term and Medium to Long-Term: From a Focus on the Short-Term to a Planning for the Medium to Long-Term

We should plan for the medium term. The overall conclusion is that a planning based on 2020 which is for the medium term will not only ease the short-term downward pressure on the economy, enhance the drive for the transformation of the development model and structural adjustments, but will also lay a solid foundation for the development in next one or two decades which are for the long-term.

We should plan for the medium term in order to solve the problems of the short-term. With a planning based on 2020, we should gain an objective understanding of the short-term pressure in the economic downturn brought about by the economic restructuring, thereby finding a new way of equitable sustainable development. Addressing short-term problems using the short-term approaches not only can hardly solve the problem, it might also lead to greater hidden risks.

The "Thirteenth Five-Year Plan" is a critical historical period in the economic transformation and upgrading. We will face some relatively small challenges in the three years from 2015 to 2017, which once pass smoothly, will bode well for the good growth momentum from 2018 to 2020. We will face a big challenge in 2020, which once pass smoothly, will lay a foundation for a 6–7% of annual economic growth for the next ten years, as there will be huge potentials for economic growth in the great market of 1.3 billion people.

3.3 Policies and Institutions: From Relying on Policy Stimulus to Focusing on Institutional Innovation

The key is to boost the vitality of the market. In the current economic situation, the short-term policy stimulus serves as the last resolve. But to solve the problem of the economic downward pressure, we should not rely mainly on policy stimulus, as the key to the solution is to appropriately deal with the relationship between the government and the market, giving full play to the decisive role of the market in resource allocation with the government playing a better role, in order to meet the development trends and realise the magnifying effect of the policies through institution innovation.

Full liberalisation of the services sector market is key. Liberalising services sector market is the key in the economic restructuring and the strategic focus of the market-oriented reform. In the 37 years since the reform and opening up, most areas in the market of the secondary industries have been highly open, but there remains a serious shortfall in the market liberalisation in the services sector, as it is still difficult to "make the cake bigger" in the services sector using domestic social funds or foreign capital. The "Thirteenth Five-Year Plan" will promote the deepening of market-oriented reform and let the market play a decisive role in allocating resources, and the top priority of the reform is to let the market play a decisive role in the services sector.

We will break the structural contradictions in the development of services sector using structural reforms. The overall conclusion is that the whole series of policies and institutional arrangements formed since the reform and opening up carry with them the salient features of incentivising the development of secondary industries and inhibiting the development of the services sector. Moving towards the economic transformation and upgrading led by the services sector, we need to accelerate in breaking the constraints caused by the structural contradictions, and speed up the structural reform in areas such as public finance, taxation, financial services and education.

We should accelerate the market liberalisation in public services through focusing on government procurement of public services. The overall conclusion is that the traditional system for providing public services based on government agencies and public institutions is unsustainable with the rapid increase in the demands for public goods by the society as a whole. We should accelerate the market liberalisation in public services through focusing on government procurement of public services, make full use of the forces in the market and the society to expand the supply of public services, and strive to make the scale of government procurement grow from 11.7% of the total fiscal expenditure in 2013 to 15–20% by 2020 and to make the share of services in the total government procurement increase from 9.4% in 2013 to about 30% by 2020.

China is at a critical period of transformation and change. We should not observe and analyse the situation of China from a short-term point of view. Instead, we should "understand the main trend and look at the big picture". China will basically achieve the economic transformation from an economy led by secondary industries to an economy led by the services sector under the "Thirteenth Five-Year Plan", which will release huge potentials for growth. This is the main trend and the big picture.

Discussant Notes 3

New Thinking on China's Economic Restructuring: A People-Oriented Structure Theory

Professor Chang Xiuze

Professor, Academy of Macroeconomic Research, National Development and Reform Commission People's Republic of China

Dear Professor Tan Khee Giap and Professor Tan Kong Yam, dear friends from Singapore and Asia-Pacific including China, good morning! I am very pleased to be invited to participate in such a high-level meeting today. The topic of this seminar is the economic issues regarding Greater China, with a particular focus on the Chinese economy under the New Normal Era. According to my own research, the major and deep issues of the Chinese economy are structural problems. So I would like to present a paper to you today, which is also the abbreviation of the content of a new book I recently published. Over the past four years, I wrote a book titled "A People-Oriented Structure Theory: New Thinking on China's Economic Restructuring". I will present the main idea of my research to you, with a presentation title of "New Thinking on China's Economic Restructuring: A People-Oriented Structure Theory". The core concept is the people-oriented structure. I will talk about four questions and briefly present to you my main ideas. I have given a copy of the book to Professor Tan Khee Giap. Please read the book (430,000 characters in length) if you are interested. Now I will present my main points.

I would like to firstly talk about the problems. I summarised and categorised the Chinese economy's structural problems into six structures. We will analyse them one by one. The first is the demand structure. The core issue of the demand structure is the lack of demand, especially the lack of consumer demand. Since 1978, the consumption share of GDP has been showing a downward trend over more than 30 years, with the latest number being about 48%. The trends of consumption and investment have formed a "seesaw" pattern. The total consumption rate is declining. As a humanist, I am more concerned about the consumption of residents. The data from the World Bank show that mainland China in 2012 had a household consumption rate of only 34.6%, which is more than 60% or even 70% in most of the developed countries and other developing countries, with the lowest among them being more than 50%. But ours is only 34.6%. The number reported by the National Bureau of Statistics of People's Republic of China is 36 %. The extremely low residents' consumption rate is a core issue.

The second is the supply structure. Vice President He Dongni just now talked about the low proportion of our services industry, which was 48.2% last year. It got a little better in the first half of this year, reaching 49.5%. But it is still less than 50%, while the world average is 70.2%. China's value added by the tertiary industry accounted for only 48.2% of the total value added of the country last year, which is lower than the world average by a difference of 22% points. The problem in our supply structure is our second problem.

The third is factor inputs structure, which decides what the economic growth of this country relies on. According to the "Global Competitiveness Report 2014–2015" released by the World Economic Forum, among the 144 economies, China's competitiveness is ranked No. 28 globally. But in the "Technological Readiness" ranking which is related to human activities in innovation, China is only ranked No. 83, which is China's weakness. For input factor, we now rely on labour, capital and land, but our technology innovation is far from enough. This is the third big contradiction.

The fourth contradiction is the serious issues in resources and the environment. In my book "On the Generalised Theory of Property Rights", I mentioned: "Rethinking the financial crisis and the environmental crisis, I found that what the Americans overdraft from is family assets, and what the Chinese overdraft from is national resources, from human resources to natural resources and environmental resources. Overdrafting from assets causes troubles for the present moment; overdrafting from resources engenders long-lasting negative effects." This is a deep-seated problem of a long-term nature.

The fifth major contradiction is the contradiction in urban and rural areas. By 2012, China's gross urbanisation rate was 52.6%, but such number was calculated based on the urban residents population (defined as population residing in urban areas for more than six months). From the perspective of human development, the actual urbanisation rate in China is only 35% (according to the population by the household registration system), which is well below the 52% of the world average. Because of the barriers entrenched by the household registration system and in particular the difference in welfare accorded by a person's status in the household registration system, more than 200 million migrant workers do not enjoy an equal level of basic public services in the city.

The sixth problem is regional issues. Looking at the main indicators reflecting the economic strength of the regions including GDP, fixed asset investment, foreign trade and local fiscal revenue, the Eastern Coastal Area is still in an absolutely dominant position. The Western China with the largest land area and the second largest population among all the regions in mainland China is still lagging behind in terms of economic development. This is a lurking inherent contradiction.

This is the first part of my speech, in which I briefly covered the six structural problems. Let's now move on from contractions and problems to talk about how to solve them. Here, I would like to propose my own idea, which is using human development as a guide, the essence of the people-oriented structure. Proposing the people-oriented structure means that the first thing

I am against is the material-oriented structure that is based on material objects, or based on GDP. The second thing I am against is government-oriented structure, or government-led economic activities. The idea of being people-oriented is the opposite of the idea of being material-oriented and government-oriented. If we cannot change from being material-oriented or government-oriented, China's economic restructuring is difficult to achieve. I will not expand on the details of this idea.

I assign three connotations to the idea of being people-oriented. Horizontally, it includes "all the people", namely the poor, the rich and the middle class, with no one excluded. Longitudinally, it includes "multiple generations", including our generation and our next generation and the generation after the next generation, seeking inter-generational balance and fairness. The third connotation is "people with multiple needs", meaning to consider the variety of needs of the people, including material needs, spiritual needs and political demands.

For our future during the "Thirteenth Five-Year Plan" and even the "Fourteenth Five-Year Plan", I personally recommend to the senior level of Chinese government that we must push for the structural transformation of China based on the idea of being people-oriented, taking this idea as a main line to follow through the six structural transformations for China in the future.

For the demand structure, we should aim to improve the "residents consumption rate" and the corresponding "civilian wealth support". Based on a comprehensive analysis of investment, export, and household income and other factors during the "Thirteen Five-Year Plan", the total consumption rate is estimated to reach about 56%, and the residents' consumption rate should be at least 43%. From 37.7% in 2014 (the latest data after adjustment) to 43%, we need to achieve roughly one percentage point increase every year, thereby releasing the consumption potential of the people and creating a new pattern of "humanised consumption".

For the supply structure (industrial structure), we should make the services sector account for more than 50% of GDP, by focusing on growing the modern services industry and strategic emerging industries that are directly related to the people. In terms of the industrial structure, we need to gradually shift from being dominated by the secondary industry to becoming a services-led economy, complementing the deficiency caused by a lagging development in the producer services industry, exploring and pioneering the "big health industry" and other new areas of life-oriented services, and developing strategic emerging industries based on the concept of "multiple generations of people".

For the factor inputs structure, we should aim at "setting free people's minds" and encouraging the widest participation of the people in innovation activities by training and nurturing hundreds of thousands of "makers", and strengthen the research and development of core technologies in key areas and brand development. We should carry out institutional innovations on intellectual property protection and its related policies, which fundamentally depends on "setting free the minds" of the innovators. After realising the potential for the new technologies to promote human development, we must also consider whether the technological advancement will enable the highly developed industrial society to tighten its "control on the people". Marcuse pointed out in his famous book "One-Dimensional Man" that "technological advancement enables the advanced industrial society's control on the people, so that it can invade people's leisure time through television, movies, radio and other media, thus occupying people's personal space...

make people satisfied with the instant gratification of their material needs and pay the price of no longer pursuing freedom and imagining another way of life".

For the structural problem in resources and the environment, we should focus on the ecological welfare and the environmental human rights, seeking the ecological welfare from technology (such as LED technology), structure (such as the relocation of Shougang Corporation and the conversion of the old factory land into cultural parks and etc.), regulations and policies (such as environmental taxes and other regulatory controls for the environment, as environmental law enforcement should be as fierce as "killer" rather than as soft as "marshmallow"), and property rights.

For the urban–rural structure, we need to focus on filling the "gap" between people from urban and rural areas. The basic idea is to make efforts to terminate the "double dual structure" that consists of the urban–rural "dual structure" and the "dual structure" within cities, realising "urban–rural integration and the integration inside the cities", thus solving the "dual structure" problems of urban and rural areas. We should realise the coexistence of rural and urban areas after meeting the precondition of "urban and rural areas being prosperous together". We should focus on solving the problem of the "dual structure" with the cities, letting migrant workers enjoy the status and benefits of city residents. We need to promote the four institutional reforms including the reform of the household registration system, the reform of the rural land system, the reform of the system for allocating public resources and the reform of the social security system.

For the regional structure, we should focus on interpersonal coordination across the regions, in order to avoid "plate group collision". In 1988, I proposed the "four types of penetration" strategy for the opening up, which was gradually adopted by the government. My favouring the "balanced development" across the regions, which can be said to be a complex of mine, is clearly visible in that strategy. I now propose the implementation of an upgraded version of the "four types of penetration" development strategy, namely the upgraded version of "along the sea", "along the river", "along the line" and "along the border".

The biggest problem constraining economic restructuring is institutional barriers. As the traditional factors including demographic dividends and resources dividends that have led the "material-oriented" economic structure are gradually subsiding, "institutional dividends" become an important driving force under the "theory of the people-oriented structure". Finally, we need the institutional support and real actions of reform, in order to overcome the "government-oriented structure" and the "material-oriented structure", pushing forward the "five-pronged" reform with a particular emphasis on anti-corruption and breaking up the entrenched vested interest groups.

Chapter 3

2015 Regional Competitiveness Analysis for Greater China

3.1 BALANCING REGIONAL GROWTH AND FACILITATING STRUCTURAL CHANGE: AN OVERVIEW

The trajectories of regional growth and development of China have experienced twists and turns since 1949 when the People's Republic was established. Since the reform and opening up of the country championed by Deng Xiaoping in the late 1970s, the direction for regional development strategy set by the central government could be mainly characterised by the policy reversal from the previously balanced but slow growth across all regions of the mainland to one that allows unbalanced regional development. Growth impetus and economic dynamism were then favoured over meticulous prevention of regional disparities (Wu 2007).

Such policy realignment in regional development in China has in general worked well in the past three decades, having long started to pay dividends in a multitude of ways that represented an uplifting experience in socioeconomic terms for all the regions in the mainland (Tan, Yuan and Yoong 2015). However, such rapid rise previously unseen in human economic development history has not arrived without its own set of problems or contradictions as they are frequently referred to in China. One of those problems related to the outlook of regional development in China is the huge regional disparities engendered as a side effect of such dynamic growth of the national economy (Tan, Yuan and Yoong 2015).

3.1.1 Regional Disparities: A Limiting Factor for Further Development

If left unchecked for too long, such regional disparities in economic development and prosperity level will constitute a serious limiting factor, creating a fearsome headwind against China's efforts in boosting its further development (Tan, Yuan and Yoong 2015). It has been documented in the literature that China would have a much lower Gini coefficient, an index measuring

Figure 3.1: Map of Greater China by Region

Source: Asia Competitiveness Institute edited from a map provided at http://bbs.eduol.cn/file.php?FileName=20128151347193336.jpg

income inequality of a society,[1] had one been able to remove the gap in the level of per capita income between Eastern Coastal Area and the rest of the regions of mainland China (Li and Xu 2008), which would exert positive impacts on both China's social fairness and its long-term stability and sustainability of growth. Lack of policy coordination across the regions has also led to the excess build-up of production capacity in some industries, depressing the profitability of the firms in those sectors and inducing the festering of some form of regional protectionism (Fan 1997).

To respond to the exigency of the issue of increasingly pronounced regional disparities across Greater China, the Asia Competitiveness Institute (ACI) at the Lee Kuan Yew School of Public Policy (LKYSPP), National University of Singapore (NUS), has launched an inaugural research undertaking to study regional competitiveness among the five regions of Greater China as shown in Figure 3.1,[2] which was published in the "2014 Annual Provincial and Regional Competitiveness Analysis for Greater China Economies: Development Strategies under the New Normal". As a

[1]The Gini coefficient is technically a measure of statistical dispersion, and it can be used to measure the degree of inequality in the distribution of income, wealth, consumption expenditure and etc. in an economy. A Gini index of 0 indicates perfect equality, while an index of 100 indicates perfect inequality.

[2]The five regions of Greater China are Eastern Coastal Area in China, Central China, Western China, Northeastern China and the region of Hong Kong, Macau and Taiwan. Eastern Coastal Area in China consists of Beijing, Fujian, Guangdong, Hainan, Hebei, Jiangsu, Shandong, Shanghai, Tianjin and Zhejiang. Central China consists of Anhui, Henan, Hubei, Hunan, Jiangxi and Shanxi. Western China consists of Chongqing, Gansu, Guangxi, Guizhou, Inner

Figure 3.2: Share of Population, Land Area, Gross Regional Domestic Product and Foreign Direct Investment by Region, 2012

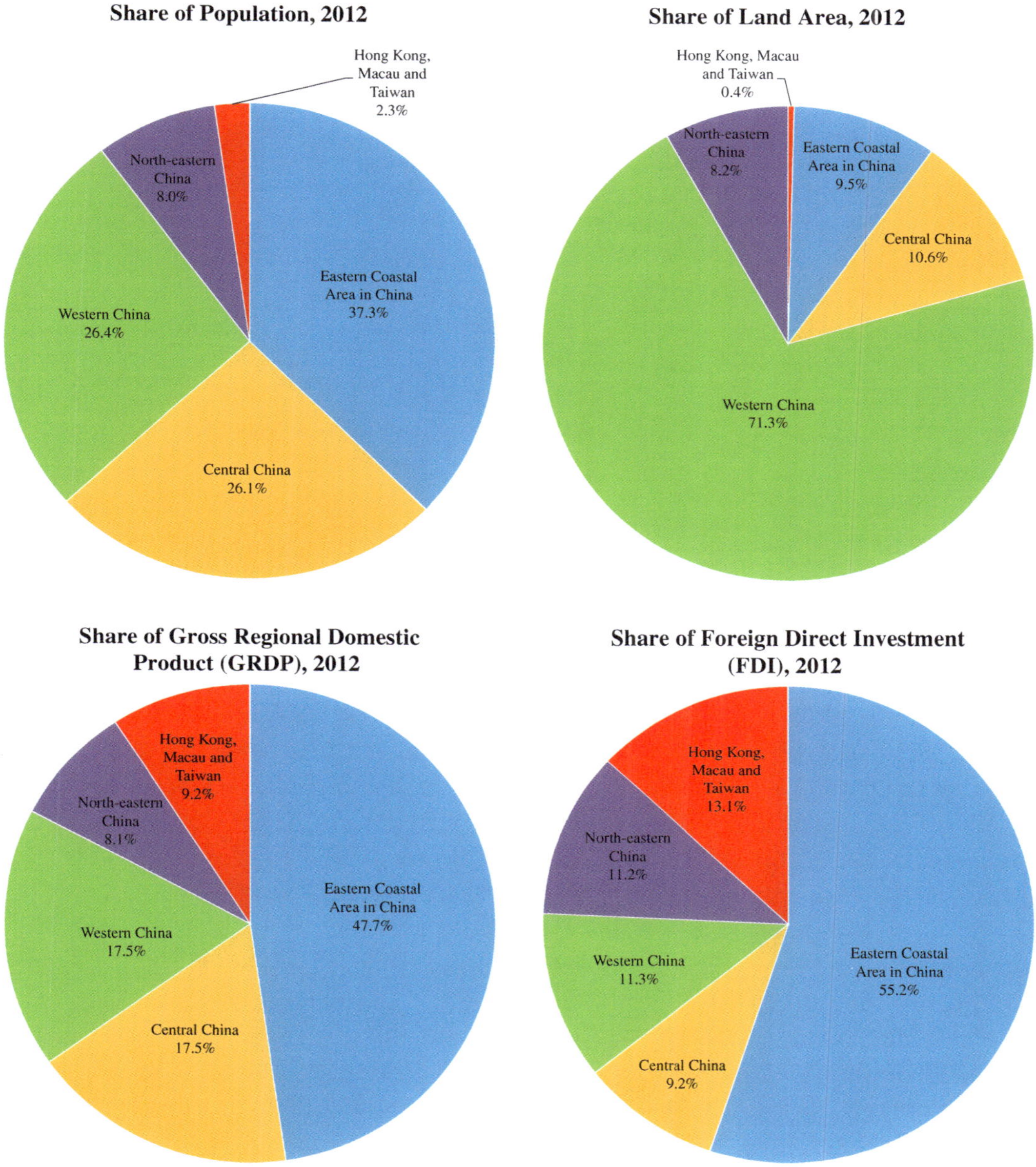

Source: Asia Competitiveness Institute based on data obtained from China Statistical Yearbook, Hong Kong Statistical Yearbook, Macau Statistical Yearbook and Taiwan Statistical Yearbook

continuation of our efforts to track and analyse the evolving competitiveness landscape at the regional level for Greater China, this chapter presents the results of our second annual regional competitiveness analysis derived from the latest data.

Mongolia, Ningxia, Qinghai, Shaanxi, Sichuan, Tibet, Xinjiang and Yunnan. North-eastern China consists of Heilongjiang, Jilin and Liaoning.

Figure 3.3: Gross Regional Domestic Product per Capita by Region, 2000 vs. 2012

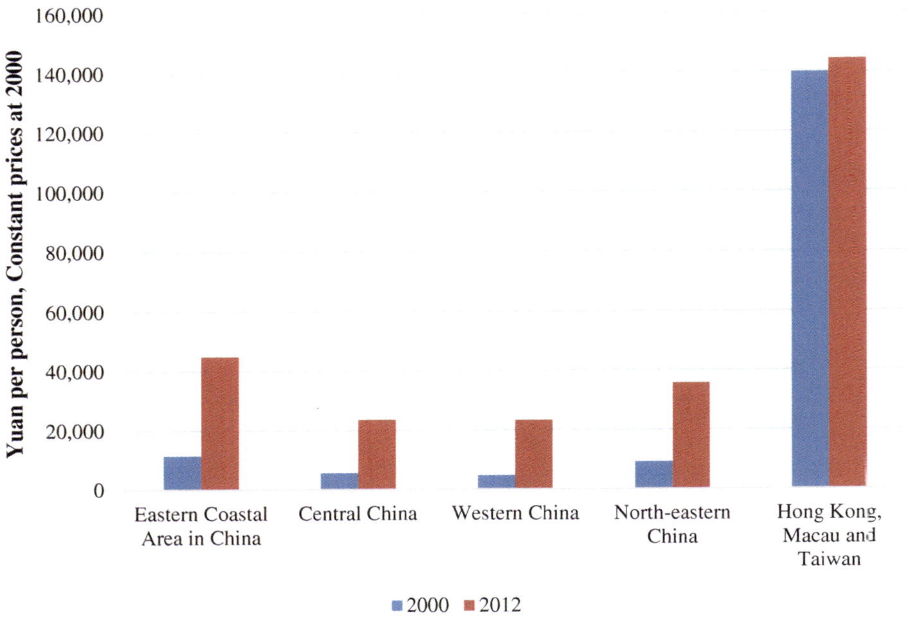

Source: Asia Competitiveness Institute based on data obtained from China Statistical Yearbook, Hong Kong Statistical Yearbook, Macau Statistical Yearbook and Taiwan Statistical Yearbook

As noted in the last volume, the over-concentration of investment[3] and development in parts of the Eastern Coastal Area has caused issues for the region itself. To name a few, environmental degradation and traffic congestion in urban areas present serious challenges negatively affecting the quality of life of the people living in this region (Chan and Yao 2008; Guo, Liu and Yu 2011). The other regions of mainland China, although having experienced unprecedented growth in the last three decades, have been lagging behind geospatially as compared to the Eastern Coastal Area (Fujita and Hu 2001).

The relatively lower level of income and standard of living, coupled with insufficient provision of quality education, healthcare and other public goods in the inland areas, has induced an exodus of the young and able from these regions into the more prosperous cities mostly located in the Eastern Coastal Area, which accounts for less than 10% of the total land area of Greater China[4] (Ha, Yi and Zhang 2015). This trend in the last 20 years has exacerbated the deficit of these regions in talents and human capital, the most vital assets required for them to catch up. Further, it has created significant social problems associated with millions of migrant workers working far away from home, as manifested in the predicament of thousands of villages whose inhabitants only consist of left-behind children and their elderly grandparents, who

[3]For example, the Eastern Coastal Area received 55.2% of the total FDI flowing into Greater China based on data in 2012, as shown in Figure 3.2.

[4]For the shares of the regions of Greater China in population, land area, economic output as measured in gross regional domestic product and FDI, please refer to Figure 3.2.

often need care themselves rather than being able to properly look after the children (Ye, Murray and Wang 2005).

These issues, which resulted from the unbalanced regional development strategy, should be deemed as problems of success rather than signs of any systemic failure of China's reform, the process of which is not without incurring mistakes that are worth reflecting upon, but the general outcome of which is an overwhelmingly constructive one.[5] Also the central government has been aware of these problems and launched targeted policy initiatives aiming to solve them since the 1990s (Zou 2008). Nonetheless, these problems, rooted in regional disparities, highlight the urgent need to spread growth centres across Greater China, failure of which will severely limit the sustainability of the much required further development of the country.

3.1.2 Rebalancing Regional Growth: A Bootstrap for Economic Restructuring

Reassuringly, plans and efforts in rebalancing regional development across China, if successfully implemented, will not only help to reduce regional disparities but will also create a domestic economic environment that is more conducive for carrying out the much required economic restructuring. Considering the exigency of the structural change needed to eradicate the root causes of the severe conundrum the excessive capacity poses to the Chinese economy as mentioned in Chapter 2, measures aimed to facilitate the process of economic restructuring directly or indirectly as in this case of pushing for a more balanced development among the regions should be going in full tilt. Indeed, rebalancing of growth across regions can go hand in hand with economic restructuring and industrial upgrading, the *condicio sine qua non* for China to transition from a middle-income country to a high-income country.

In Chapter 2 of the "2014 Annual Provincial and Regional Competitiveness Analysis for Greater China Economies: Development Strategies under the New Normal", the second instalment of our annual tracking and analysis of competitiveness for Greater China, the authors enumerated three most pressing issues China faces and needs to find solutions to in the short-to-medium time horizon in order to continue with its further development in a more sustainable manner. They are environmental degradation, income or wealth inequality and the lack of a well-functioning social security system (Tan, Yuan and Yoong 2015).

While environmental degradation is unfortunately often the by-product of economic growth especially in the course of a poor country trying to get rich, it has by now reached unbearable levels in many parts of China. Meanwhile, the latter two issues on inequality and social security were observed to be mutually reinforcing in a vicious cycle (Tan, Yuan and Yoong 2015). The lack of a comprehensive and well-functioning social security system puts in an untenable position the attempt to help ease the divide between the haves and have-nots, while socioeconomic equality, which is already alarming but still without clear signs of turning better, would erode the

[5]For example, as can be seen in Figure 3.3, there have been great increases in the annual economic output per capita in all regions of Greater China from 2000 to 2012, except for the region of Hong Kong, Macau and Taiwan, which has only experienced marginal growth in economic output per capita for the same period.

social fabric and adversely impact the confidence of the general public in the government's resolution to always put people's interests above all else, thereby making it even more difficult to implement the necessary reform measures to improve the functioning of the national socioeconomic systems including the social security system, as a weakening trust could result in lower compliance with rules and regulations set by government (OECD 2013).

As a carbuncle neglected could become the bane of one's life, if no effective counter-measures were taken to break such vicious cycle by reducing income inequality and improving social security, there would be a likely consequence that the ensuing entrenched chasm between rich and poor not only will stifle the healthy level of social mobility required for a dynamic and progressive society, but it will also worsen the conditions in which China has to cope with excessive capacity and add difficulty to the already challenging undertaking of implementing structural change, as discussed in Chapter 2.

More balanced development across regions, however, will help to alleviate all three problems (Tan, Yuan and Yoong 2015). More growth centres widely distributed across all regions of Greater China will considerably reduce the intensity with which economic activities are taxing the environment as compared to the stranded situation now we encounter where industries accountable for almost half of the total economic output of Greater China have to be narrowly squeezed into less than one-tenth of the total land area of the country.[6]

Rebalancing regional development will also make significant contributions in decreasing national income inequality and reduce the Gini coefficient of the country (Fan and Sun 2008), which helps in improving social fairness and nurturing a harmonious society. Furthermore, more balanced growth among the regions of Greater China will render it more viable for the central government to build a nationally integrated social security system incorporating both rural and urban areas of the country, which will be much more cost effective by pooling substantially more risks together and in so doing be more comprehensive and welfare enhancing for the people.

The fruits of rebalanced regional development across Greater China will in turn boost the public confidence, strengthen the allegiance of the people and cohesiveness of the society, thus increasing social stability, a prerequisite for any nation to achieve long-term prosperity, as concurred by Li Yining who proclaims that "harmony of the society is the biggest institutional dividend" for China (Li 2014). Thus, rebalancing regional growth will come to the aid of pushing forward the critical economic restructuring by helping to curtail the vitiating factors such as the widening gap between the rich and the poor and to foster positive public opinions and people's willingness to support the reform policies. In due course, a more balanced growth pattern across China will also boost effective demands, thereby lessening the burden caused by excessive capacity.

There are encouraging signs that the invisible hand of the market economy has already shown its constructive effects in regional rebalancing. The so-called domestication of the Flying Geese Model[7] has been taking place in recent years where labour-intensive manufacturing

[6]For the shares of the regions of Greater China in population, land area, economic output as measured in gross regional domestic product and FDI, please refer to Figure 3.2.

[7]The Flying Geese Model was first proposed by the Japanese scholar Kaname Akamatsu to explain the patterns of international division of labour and transfer of technology among the different countries and especially those in East

industries are increasingly being relocated towards the inland regions of Greater China (Tan, Yuan and Yoong 2015). At the same time, the coastal cities are upgrading their industries with a focus on building their strength in services sectors (Fang, Wang and Yue 2009; Tan *et al.* 2014).

So the visible hand of the government is not working alone in this case, and the role of the government should focus on the creation of a strategic platform upon which such benign forces of the market can be harnessed to help diversify the distribution of industries in line with the comparative advantages of each region and spread growth poles all across Greater China. Such relocation and upgrading of industries within and among the regions will also induce increasingly deeper division of labour and therefore further efficiency enhancement (Kojima 2000; Fang, Wang and Yue 2009), the very essence China needs to sustain its growth before it achieves the high-income status.

That said, one should not get carried away when plotting the strategies for regional growth rebalancing. Special attention should at least be paid to two points. First, when relocating industries to inland areas in central, western and northeastern parts of China, we should avoid the reluctant mentality towards environmental protection so as not to repeat the inappropriate paradigm of "pollute first, control later". The accumulated debts resulted from the negligence in caring for the environment have reached a critical level, ticking like time bombs in various areas of the country, if not already burst out with catastrophic consequences.

When developing the less advanced inland areas, we are actually blessed with the so-called latecomer advantage (Veblen 1915; Lin 2011), which is analogous to the domestication of the Flying Geese Model in a national context. More up to date knowledge and advanced technologies together with more experiences and lessons distilled from history are available to us, and it is up to the decision makers of the day to conscientiously adopt them in order to prepare for a harvest in future that is beneficial to the vast majority of the people, instead of only enriching some minorities.

Second, as one should not strive for absolute equality in income or wealth distribution among the people because that will almost invariably lead to degeneration of incentives for individuals where stagnation of the economy and suffocation of entrepreneurship will grip the nation before long, we should not target absolutely equal paces of development and levels of prosperity across the regions. As China is still a developing country with considerable gaps *vis-à-vis* the advanced countries in the West in terms of many per capita measures in income and productivity levels,[8] maintaining a healthy level of growth momentum that is sufficient to provide enough employment creation and improve the overall efficiency of the economy should still remain the primary concern in its national economic strategy.

Measures aimed at rebalancing regional development should be designed and implemented to facilitate the robustness, efficiency and sustainability of the national economy, but should not be

---

Asia based on dynamic comparative advantage. For more explanation and details about this model, please refer to "The Flying Geese Model of Asian Economic Development: Origin, Theoretical Extensions and Regional Policy Implications" by Kiyoshi Kojima published in the *Journal of Asian Economics* (Kojima 2000).

[8]For example, according to the data from the World Bank, in terms of the gross national income (GNI) per capita, China is about 22% of that of the United States (US) in 2013 even after accounting for purchasing power parity.

executed in an isolated mode for its own sake. As achieving true prosperity for China is a long-term game, evening up the level of affluence and standard of living across all the regions to a desirable degree requires lasting efforts as pronounced regional disparities are a widespread phenomenon among the countries in the world including the most advanced ones.[9]

As alluded to in Chapter 2, the newly established institutions such as the Asian Infrastructure Investment Bank (AIIB) and initiatives like "One Belt, One Road" (OBOR) may help to create a favourable external environment to ease the syndrome of excessive capacity for China through constructing win–win partnerships around the globe. A more balanced growth pattern across the regions in China will also go a long way in nurturing the right internal or domestic conditions for more effective utilisation of factors of production, thereby bridging the gap between effective demands and supplies and alleviating excessive capacity.

In light of the domestication of the "Flying Geese", given the geographic vastness and diversity of China, rebalancing regional growth can indeed be considered as an integral part of implementing OBOR and other large-scale economic initiatives involving partnerships between China and other countries, which will bring about great opportunities for facilitating the development across all regions of Greater China, as conditions in favour of a rebalanced economic development in Greater China have been created against such a backdrop and the strategic role of inland regions of Greater China will feature much more prominently than in previous decades.

With a greater fiscal space made available by these initiatives, we are confident that much can be achieved by various levels of government in Greater China under the New Normal Era for the whole country to embark on a more sustainable trajectory of development with better quality of growth and to make benefits of such growth more widely shared by the people. Therefore, we hope the research results presented in the subsequent sections of this chapter and the following chapters of the book can serve as a modest spur to induce intellectual debate and further research, contributing instructive insights for drawing up astute policy measures that will effectively facilitate the rebalancing of development in the regions of Greater China as well as the economic structuring required for resolving the excessive capacity, in order for the country to eventually transit from a developing country to a developed one.[10]

For compatibility with the official sources of data from which many indicators used in our competitiveness analysis are derived, we adopt the demarcation by the National Bureau of Statistics of China in terms of the four regions constituting the mainland of Greater China, which are the Eastern Coastal Area in China, Central China, Western China and North-eastern China, as illustrated in Figure 3.1. Considering the special relationship between mainland China and

[9]Even countries such as the US, Britain and France have not achieved a homogenous level of development across the regions within their own borders. Please refer to the report by *The Economist* regarding the existing gap between the rich and poor regions in these three developed countries, a situation only exacerbated by the Global Financial Crisis, available at http://www.economist.com/node/18332880.

[10]The results of our research on regional competitiveness analysis for Greater China has recently been cited in the public policy discourse in China. See, for example, in Chapter 1 of "Understand One Belt, One Road", a book authored by economic advisors to the State Council and top think tank economists from China (Li *et al.* 2015).

Hong Kong, Macau and Taiwan and the similarities among these three Greater China economies in terms of the level of economic development and institutional setup, they are grouped as another region of Greater China in our study.

The rest of Chapter 3 is organised as follows. Section 3.2 presents the research findings which include the Overall Competitiveness ranking as well as the competitiveness rankings of four environments. Besides providing a snapshot of the 2015 ACI rankings, this section will also track the changes of each region in terms of competitiveness over time. Details of the sub-environments under each environment will also be discussed. Section 3.3 concludes by summarising salient results of the study.

3.2 RESEARCH FINDINGS

This section presents the 2015 competitiveness rankings of the five regions in Greater China, which were derived according to the methodological framework discussed in Chapter 1 of this book. Similar to the competitiveness study on 34 Greater China economies, the Overall Competitiveness ranking as well as the rankings for the four environments are generated based on the standardised scores of each region.

3.2.1 ACI 2015 Overall Competitiveness Rankings and Scores

Table 3.1 shows the Overall Competitiveness ranking and scores for the five regions in Greater China. It is unsurprising that the Eastern Coastal Area in China attains the top position with a score of 1.7618, as six out of the 10 economies of the Eastern Coastal Area are among the top 10 in competitiveness ranking of the 34 Greater China economies (see Table 2.1 of Chapter 2). In contrast, Central China is ranked last with a score of –0.8960.

The average performing region (with a standardised score of 0) is somewhere between the #2-ranked Hong Kong, Macau and Taiwan (0.4577) and the #3-ranked Western China (–0.5982). The score of Western China is the median score for Greater China, as it is in the middle of the ranking (#3). Clearly, we can see that the average score (0) is slightly higher than the median score (–0.5982). The average value is positioned in the upper half of the ranking, above the centre of the list.

The spread of the Overall Competitiveness scores in the regional ranking is not as wide as in the ranking for individual economies. The difference in the scores between the top-performing region and the worst-performing region is 2.6578, as compared to 3.7156 which is the difference in the scores between the #1-ranked economy and the #34-ranked economy. The largest gap in the regional scores is between the Eastern Coastal Area in China and Hong Kong, Macau and Taiwan (1.3041[11]), followed by that between Hong Kong, Macau and Taiwan and Western China

[11]The gap in the scores between the Eastern Coastal Area in China and Hong Kong, Macau and Taiwan is calculated as follows: Score of the Eastern Coastal Area in China — Score of Hong Kong, Macau and Taiwan = 1.7618 –0.4577 = 1.3041.

Table 3.1: ACI 2015 Overall Competitiveness Ranking and Scores[12]

Rank	Region	Score
1	Eastern Coastal Area in China	1.7618
2	Hong Kong, Macau and Taiwan	0.4577
3	Western China	–0.5982
4	North-eastern China	–0.7253
5	Central China	–0.8960

Source: Asia Competitiveness Institute

(1.0559). These sizeable gaps in the Overall Competitiveness scores, as evident in the Eastern Coastal Area in China being almost four times more competitive than Hong Kong, Macau and Taiwan, are striking because the two regions are only one rank apart. The scores of the Western, Northeastern China and Central China regions are rather close, portending some rivalry as a healthy way to strengthen Overall Competitiveness of China as a whole.

We can also look at the rankings of the five regions in Greater China from a geographical perspective. In Figure 3.4, regions are colour-coded according to their level of competitiveness; the regions that are more competitive are shaded darker while the less competitive regions have lighter colours.

Table 3.2 shows the Overall Competitiveness rankings and scores of the five regions in Greater China from 2000–2012. The Eastern Coastal Area has surpassed Hong Kong, Macau and Taiwan as #1 since 2002 with its score steadily increasing from 1.0265 in 2000 to 1.7618 in 2012. In contrast, the score of Hong Kong, Macau and Taiwan has been declining over the same period. Since the Eastern Coastal Area has assumed the top position in 2002, the gap in the scores between the #1-ranked and the #2-ranked regions has been expanding, from a difference of 0.1239 in 2002 to 1.3041 in 2012.

The scores of Central, Western and North-eastern China have remained negative over the past decade (2000–2012), implying that they have below average performance among all regions of Greater China. It is heartening to see the progressive improvement in the competitiveness of Western China as it has improved its score from –0.8002 in 2000 to –0.5982 in 2012. In 2007 and since 2010, Western China has replaced Northeastern China as the #3 in Greater China region. At the same time, North-eastern China slipped from its #3 position to #4 in 2007 and since 2010. In terms of Overall Competitiveness, Central China has consistently ranked last. However, there is a slight improvement in the scores of Central China.

In order to further analyse the competitiveness over time of each region, we split the 13-year period into two periods. This offers a better comparison of improvements in competitiveness between 2000–2005 (Period A) and 2006–2012 (Period B). In Figure 3.5, the horizontal axis

[12]Result shown in Table 3.1 is based on 2012 data, collated and analysed in 2015.

Figure 3.4: ACI 2015 Overall Competitiveness Ranking Map[13]

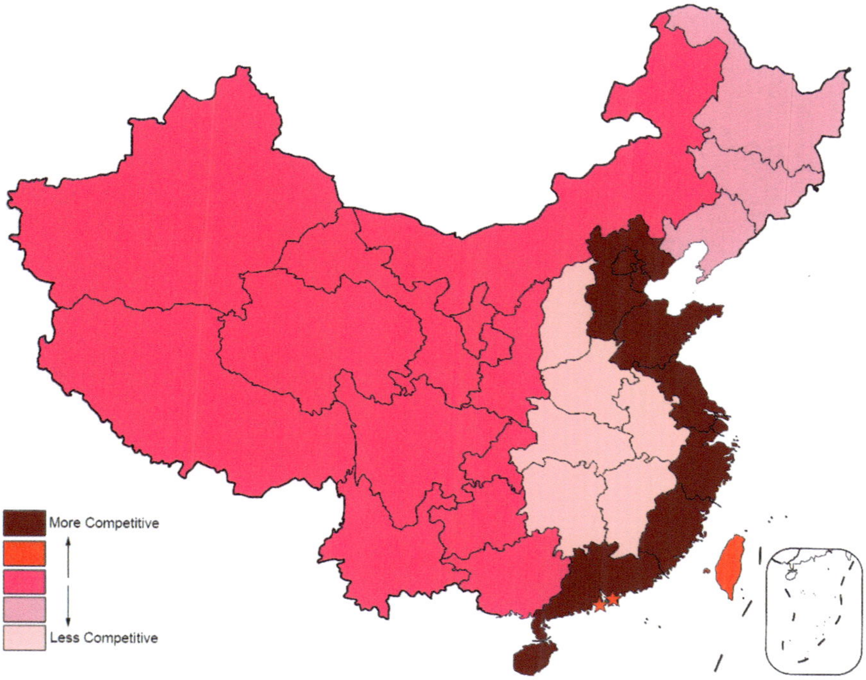

Source: Asia Competitiveness Institute edited from a map provided at http://bbs.eduol.cn/file.php?FileName=20128151347193336.jpg

shows the average Overall Competitiveness score over the period of 2000–2005 for each region and the vertical axis records the average score for 2006–2012. We observe that the Eastern Coastal Area, Western China and Central China are above the 45-degree line, while Hong Kong, Macau and Taiwan as well as North-eastern China are below the 45-degree line. Those regions which are above the 45-degree line performed better in Period B as compared to the earlier Period A, that is, their Overall Competitiveness has improved over time. On the other hand, Hong Kong, Macau and Taiwan as well as North-eastern China fared better in Period A.

The improvement of the Eastern Coastal Area is the most remarkable among all regions in Greater China as seen in the large vertical distance between the blue diamond and the 45-degree line. Since North-eastern, Western and Central China cluster closely along the 45-degree line, their changes in competitiveness between the two periods are marginal. Hong Kong, Macau and Taiwan, in contrast, has seen sharp deterioration in Overall Competitiveness over these two periods, as evidenced by the significant vertical distance between the blue cross and the 45-degree line.

---

[13]Result shown in Figure 3.4 is based on 2012 data, collated and analysed in 2015.

Table 3.2: ACI 2015 Overall Competitiveness Rankings and Scores: Evolution over Time[14]

	1		2		3	
Rank	**Region**	**Score**	**Region**	**Score**	**Region**	**Score**
Overall Competitiveness						
2000	Hong Kong, Macau and Taiwan	1.3787	Eastern Coastal Area in China	1.0265	North-eastern China	–0.5603
2001	Hong Kong, Macau and Taiwan	1.2914	Eastern Coastal Area in China	1.1446	North-eastern China	–0.6628
2002	Eastern Coastal Area in China	1.2793	Hong Kong, Macau and Taiwan	1.1554	North-eastern China	–0.6314
2003	Eastern Coastal Area in China	1.2859	Hong Kong, Macau and Taiwan	1.1452	North-eastern China	–0.6587
2004	Eastern Coastal Area in China	1.3384	Hong Kong, Macau and Taiwan	1.0896	North-eastern China	–0.7071
2005	Eastern Coastal Area in China	1.4746	Hong Kong, Macau and Taiwan	0.9139	North-eastern China	–0.6347
2006	Eastern Coastal Area in China	1.4899	Hong Kong, Macau and Taiwan	0.9100	North-eastern China	–0.6949
2007	Eastern Coastal Area in China	1.5034	Hong Kong, Macau and Taiwan	0.8947	Western China	–0.7349
2008	Eastern Coastal Area in China	1.6031	Hong Kong, Macau and Taiwan	0.7467	North-eastern China	–0.6551
2009	Eastern Coastal Area in China	1.6917	Hong Kong, Macau and Taiwan	0.5985	North-eastern China	–0.6356
2010	Eastern Coastal Area in China	1.7377	Hong Kong, Macau and Taiwan	0.5156	Western China	–0.6847
2011	Eastern Coastal Area in China	1.7508	Hong Kong, Macau and Taiwan	0.4808	Western China	–0.6244
2012	Eastern Coastal Area in China	1.7618	Hong Kong, Macau and Taiwan	0.4577	Western China	–0.5982

[14]Result shown in Table 3.2 is based on 2000–2012 data, collated and analysed in 2015.

	4		5	
Rank	**Region**	**Score**	**Region**	**Score**
2000	Western China	–0.8002	Central China	–1.0446
2001	Western China	–0.8136	Central China	–0.9596
2002	Western China	–0.8589	Central China	–0.9444
2003	Western China	–0.7636	Central China	–1.0088
2004	Western China	–0.7179	Central China	–1.0030
2005	Western China	–0.7193	Central China	–1.0344
2006	Western China	–0.7640	Central China	–0.9410
2007	North-eastern China	–0.7427	Central China	–0.9206
2008	Western China	–0.7853	Central China	–0.9094
2009	Western China	–0.7678	Central China	–0.8869
2010	North-eastern China	–0.6963	Central China	–0.8723
2011	North-eastern China	–0.6980	Central China	–0.9091
2012	North-eastern China	–0.7253	Central China	–0.8960

Source: Asia Competitiveness Institute

Figure 3.5: ACI 2015 Competitiveness Over Time for Overall Competitiveness[15]

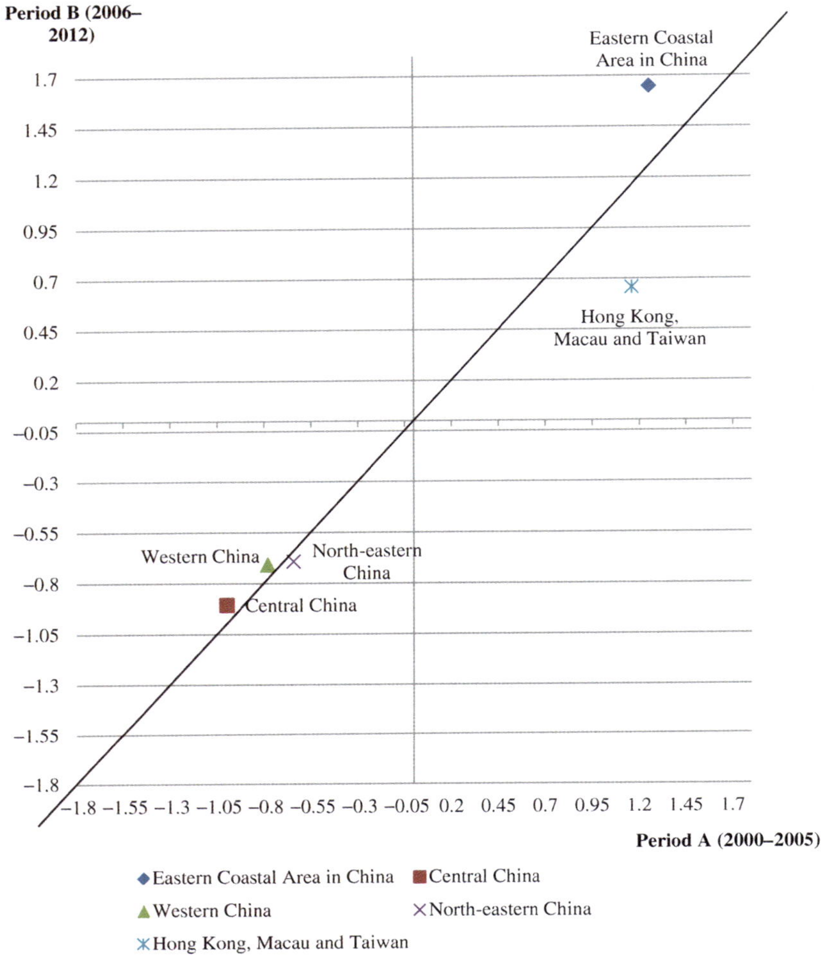

Source: Asia Competitiveness Institute

3.2.2 ACI 2015 Macroeconomic Stability Rankings and Scores

Table 3.3 shows the ranking and scores for the first environment, Macroeconomic Stability, for the five regions in Greater China. Similar to the Overall Competitiveness ranking, the Eastern Coastal Area in China tops the chart with a score of 1.9748. However, Central China no longer occupies the last position, North-eastern China is ranked last instead.

The average performing region (with a standardised score of 0) in the Macroeconomic Stability ranking is somewhere between the #1-ranked Eastern Coastal Area in China (1.9748) and the #2-ranked Hong Kong, Macau and Taiwan (–0.2706). The region with the median score is Western China. Again, the average score (0) of the ranking is higher than the median score (–0.4430).

It is worth noting that there is a marked drop in the scores from the #1-ranked Eastern Coastal Area (1.9748) to the #2-ranked Hong Kong, Macau and Taiwan (–0.2706). The gap in the scores

[15] Result shown in Figure 3.5 is based on 2000–2012 data, collated and analysed in 2015.

Table 3.3: ACI 2015 Macroeconomic Stability Ranking and Scores[16]

Rank	Region	Score
1	Eastern Coastal Area in China	1.9748
2	Hong Kong, Macau and Taiwan	–0.2706
3	Western China	–0.4430
4	Central China	–0.4974
5	North-eastern China	–0.7638

Source: Asia Competitiveness Institute

Figure 3.6: ACI 2015 Macroeconomic Stability Ranking Map[17]

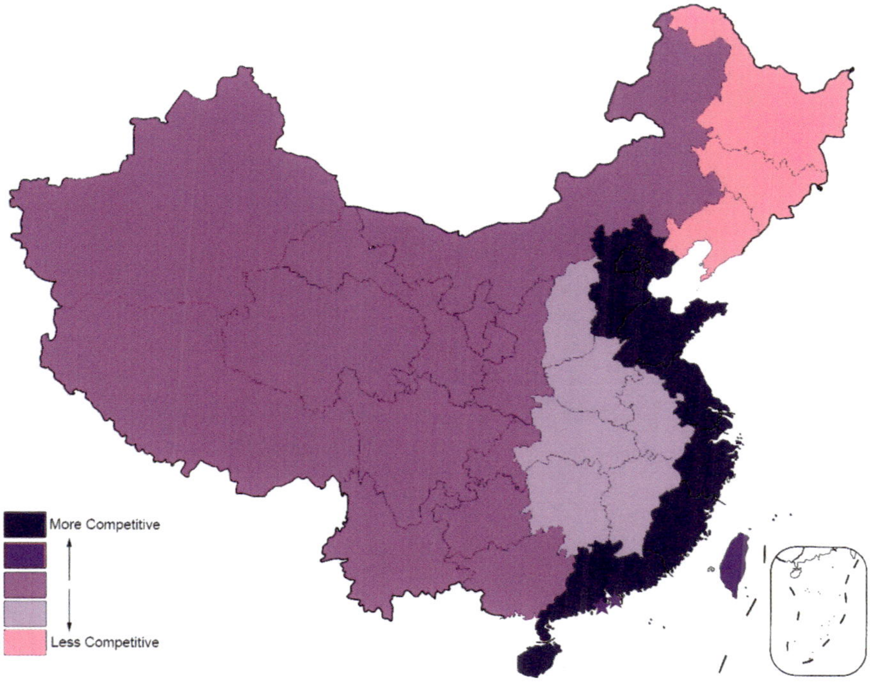

Source: Asia Competitiveness Institute edited from a map provided at http://bbs.eduol.cn/file.php?FileName=20128151347193336.jpg

between these two regions is 2.2454. This means that the Eastern Coastal Area has considerably outperformed the remaining regions in Greater China in terms of Macroeconomic Stability.

Figure 3.6 shows the Macroeconomic Stability ranking of the five regions in Greater China from a geographical perspective. Regions are colour-coded according to their level of competitiveness, whereby more competitive regions are shaded darker, while the less competitive regions have lighter colours.

Table 3.4 shows the Macroeconomic Stability rankings and scores for the five regions of Greater China from 2000–2012. The Eastern Coastal Area in China as well as Hong Kong,

[16]Result shown in Table 3.3 is based on 2012 data, collated and analysed in 2015.

[17]Result shown in Figure 3.6 is based on 2012 data, collated and analysed in 2015.

Table 3.4: ACI 2015 Macroeconomic Stability Rankings and Scores: Evolution over Time[18]

	1		2		3	
Rank	**Region**	**Score**	**Region**	**Score**	**Region**	**Score**
Macroeconomic Stability						
2000	Eastern Coastal Area in China	1.6985	Hong Kong, Macau and Taiwan	0.5759	North-eastern China	–0.6121
2001	Eastern Coastal Area in China	1.6892	Hong Kong, Macau and Taiwan	0.6104	Western China	–0.6819
2002	Eastern Coastal Area in China	1.8462	Hong Kong, Macau and Taiwan	0.2586	North-eastern China	–0.5601
2003	Eastern Coastal Area in China	1.8133	Hong Kong, Macau and Taiwan	0.3444	Western China	–0.5948
2004	Eastern Coastal Area in China	1.8141	Hong Kong, Macau and Taiwan	0.3501	Western China	–0.5970
2005	Eastern Coastal Area in China	1.9237	Hong Kong, Macau and Taiwan	–0.0537	North-eastern China	–0.4668
2006	Eastern Coastal Area in China	1.9364	Hong Kong, Macau and Taiwan	–0.0002	Central China	–0.6279
2007	Eastern Coastal Area in China	1.9246	Hong Kong, Macau and Taiwan	0.0453	North-eastern China	–0.6426
2008	Eastern Coastal Area in China	1.9365	Hong Kong, Macau and Taiwan	–0.0456	North-eastern China	–0.4436
2009	Eastern Coastal Area in China	1.9829	Hong Kong, Macau and Taiwan	–0.2512	Central China	–0.5293
2010	Eastern Coastal Area in China	1.9794	Hong Kong, Macau and Taiwan	–0.2321	Western China	–0.5247
2011	Eastern Coastal Area in China	1.9926	Western China	–0.3981	Hong Kong, Macau and Taiwan	–0.4237
2012	Eastern Coastal Area in China	1.9748	Hong Kong, Macau and Taiwan	–0.2706	Western China	–0.4430

[18]Result shown in Table 3.4 is based on 2000–2012 data, collated and analysed in 2015.

	4		5	
Rank	**Region**	**Score**	**Region**	**Score**
2000	Western China	–0.7142	Central China	–0.9480
2001	Central China	–0.7938	North-eastern China	–0.8239
2002	Western China	–0.6776	Central China	–0.8671
2003	North-eastern China	–0.6862	Central China	–0.8768
2004	North-eastern China	–0.7499	Central China	–0.8172
2005	Western China	–0.4848	Central China	–0.9184
2006	North-eastern China	–0.6398	Western China	–0.6685
2007	Central China	–0.6612	Western China	–0.6660
2008	Western China	–0.6814	Central China	–0.7659
2009	Western China	–0.5713	North-eastern China	–0.6312
2010	Central China	–0.5589	North-eastern China	–0.6637
2011	Central China	–0.5304	North-eastern China	–0.6404
2012	Central China	–0.4974	North-eastern China	–0.7638

Source: Asia Competitiveness Institute

Macau and Taiwan have been consistently ranked #1 and #2, respectively across the years (2000–2012), except in 2011 with Hong Kong, Macau and Taiwan briefly overtaken by Western China. The gap in competitiveness between the top and second positions has widened over the years. While the score of the Eastern Coastal Area has increased from 1.6985 in 2000 to 1.9748 in 2012, the score of Hong Kong, Macau and Taiwan has deteriorated from 0.5759 in 2000 to –0.2706 in 2012. This has caused the difference in the scores between the #1-ranked region and the #2-ranked region to expand from 1.1226 in 2000 to 2.2454 in 2012. During the same period, the rankings of Central, Western and North-eastern China have fluctuated between #3 and #5 with no distinct pattern, with the exception in 2011 when Western China briefly ranked at #2.

The Macroeconomic Stability environment is made up of three sub-environments, namely Regional Economic Vibrancy; Openness to Trade and Services; and Attractiveness to Foreign Investors. Through Tables 3.3 and 3.4 as well as Figure 3.6, we learn that the Eastern Coastal Area is the most competitive in terms of Macroeconomic Stability, followed by Hong Kong, Macau and Taiwan. Figure 3.7 provides more details about whether each region performs particularly well or poorly and in each sub-environment.

Although both the Eastern Coastal Area and Hong Kong, Macau and Taiwan excel in Openness to Trade and Services, as shown in their above-zero scores in this sub-environment, Regional Economic Vibrancy and Attractiveness to Foreign Investors of Hong Kong, Macau and Taiwan are alarmingly poor. This should serve as a reminder for the region to work on improving its competitiveness performance in this environment.

As for Central, Western and North-eastern China, they perform rather weakly in terms of Openness to Trade and Services as well as Attractiveness to Foreign Investors. That said, North-eastern China has an additional sub-environment to work upon which is Regional Economic Vibrancy.

Figure 3.8 illustrates competitiveness over time for Regional Economic Vibrancy. The horizontal axis of the graph shows average scores for each region for the period 2000–2005 and the vertical axis records those during 2006–2012. The blue diamond dot denoting the Eastern

Figure 3.7: ACI 2015 Sub-environment Scores of Macroeconomic Stability[19]

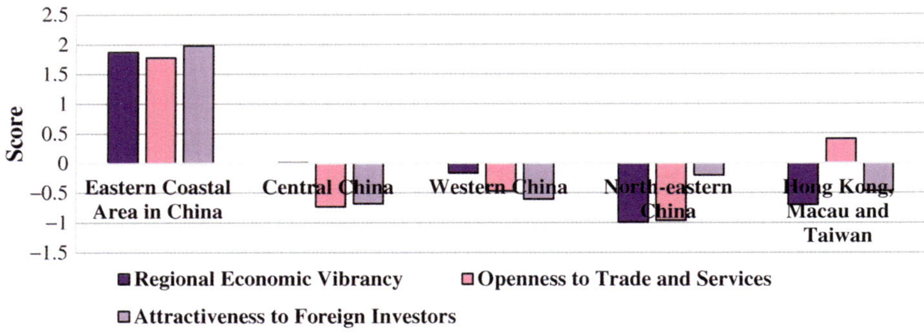

Source: Asia Competitiveness Institute

[19]Results shown in Figures 3.7–3.10 are based on 2012 data, collated and analysed in 2015.

Figure 3.8: ACI 2015 Competitiveness Over Time for Regional Economic Vibrancy

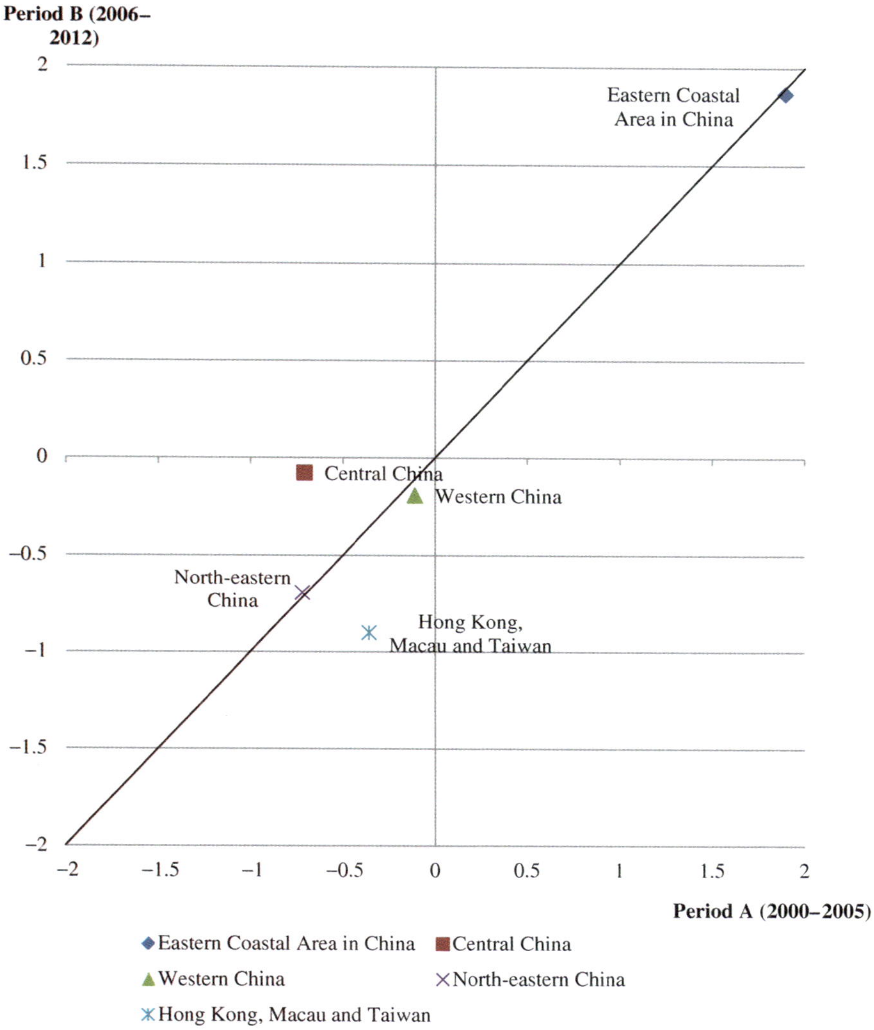

Source: Asia Competitiveness Institute

Coastal Area in China falls on the top right corner of Figure 3.8, almost exactly on the 45-degree line. This shows that the average score of the region remained roughly the same for the two periods 2000–2005 and 2006–2012. Not only is the region noticeably more competitive than other regions in this particular sub-environment (as shown in its position on the top right corner of Figure 3.8, far away from the rest of the dots), the competitiveness of this region has remained unchanged over time.

The red square denoting Central China lies in the third quadrant above the 45-degree line. It shows that this region performed better in the later period, which means that its competitiveness in this sub-environment has improved over time. In Period A (2000–2005), the average Regional

Economic Vibrancy score for the region is –0.7080 and it has increased to –0.0772 in Period B (2006–2012). We can see that the Central China is also not as competitive as the Eastern Coastal Area (which lies at the top right corner of Figure 3.8) in this particular sub-environment.

The green triangle denoting Western China falls in the middle of Figure 3.8 slightly below the 45-degree line. This shows that the average score of the region deteriorated slightly over the two periods 2000–2005 and 2006–2012. Western China is apparently less competitive than the Eastern Coastal Area in China which is at the top right corner of Figure 3.8.

The black cross denoting the North-eastern China falls on the bottom left corner of Figure 3.8 slightly above the 45-degree line. It shows that this region has been doing slightly better in the later period as its competitiveness in this sub-environment has improved only marginally over time. In Period A (2000–2005), the average Regional Economic Vibrancy score for the region is –0.7198, which is lower than its average score of –0.6947 in Period B (2006–2012).

The blue asterisk denoting Hong Kong, Macau and Taiwan falls in the third quadrant of Figure 3.8, far below the 45-degree line. This shows that the average score of the region is much higher in Period A (2000–2005) as compared to Period B (2006–2012). Not only is the region noticeably less competitive than the Eastern Coastal Area in China (as shown in the Eastern Coastal Area's position on the top right corner of Figure 3.8 far away from the rest of the regions) in this particular sub-environment, but the competitiveness of this region has diminished significantly over time.

In Figure 3.9 pertaining to Openness to Trade and Services, the blue diamond dot denoting the Eastern Coastal Area in China is above the 45-degree line. It shows that this region performed better in the later period as its competitiveness in this sub-environment has improved over time. In Period A (2000–2005), the average Openness to Trade and Services score for the region is 0.9875, while in Period B (2006–2012), its average score is 1.5315.

The red square representing Central China falls on the lower left corner of Figure 3.9, almost exactly on the 45-degree line. This shows that Central China is the least competitive in terms of Openness to Trade and Services as compared to the other regions in Greater China. Also, its competitiveness has remained largely unchanged between the two periods; the average scores of Central China in these two periods are almost the same.

The green triangle representing Western China lies at the lower left corner of Figure 3.9, slightly above the 45-degree line. This shows that the competitiveness of Western China has improved over time as its average score of –0.7994 in Period A (2000–2005) is lower than its average score of –0.6879 in Period B (2006–2012).

The black cross representing North-eastern China falls on the lower left corner of Figure 3.9, slightly below the 45-degree line. This shows that North-eastern China has performed better in the earlier period. The average score of North-eastern China in Period A (2000–2005) is –0.7606 as compared to –0.8072 in Period B (2006–2012).

In Figure 3.9, the blue asterisk denoting Hong Kong, Macau and Taiwan is also far below the 45-degree line. It shows that this region performed better in the earlier period as its competitiveness in this sub-environment has declined considerably over time. In Period A (2000–2005), the average Openness to Trade and Services score for the region is 1.4224, while in Period B (2006–2012), its average score is 0.8214.

Figure 3.9: ACI 2015 Competitiveness Over Time for Openness to Trade and Services

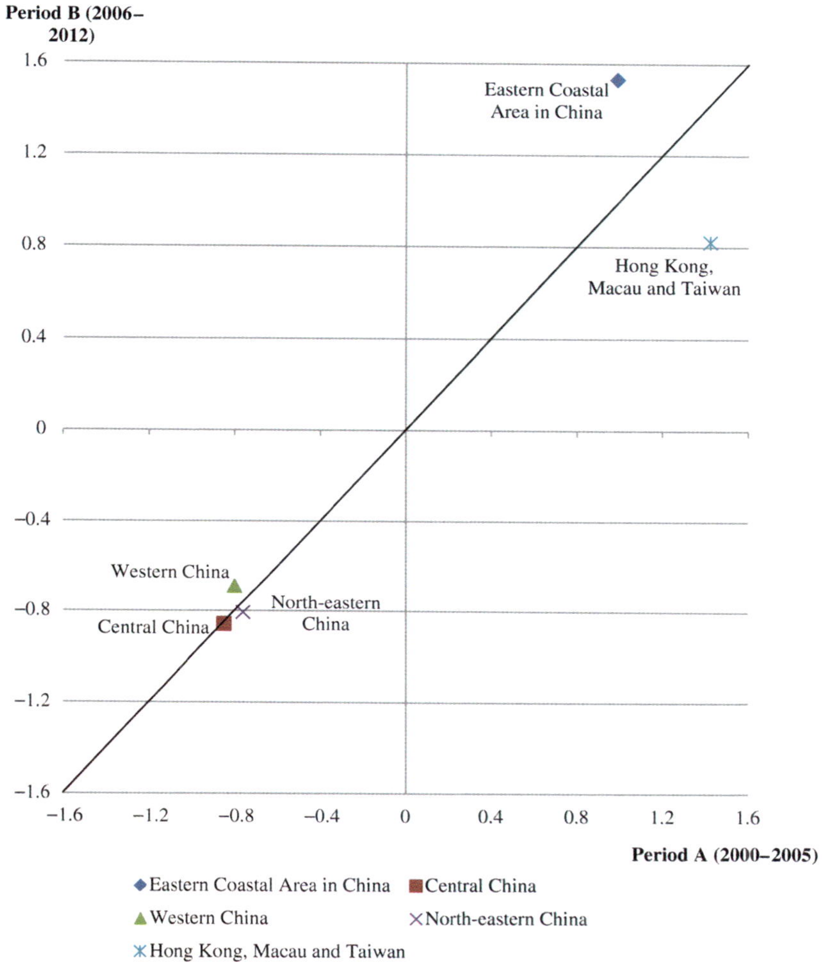

Source: Asia Competitiveness Institute

Figure 3.10 on Attractiveness to Foreign Investors shows that the Eastern Coastal Area in China is slightly above the 45-degree line. It shows that the region has improved in terms of competitiveness in the later period as its average score increased from 1.9015 in Period A (2000–2005) to 1.9636 in Period B (2006–2012).

Figure 3.10 also shows Central China lying in the third quadrant, slightly above the 45-degree line. It shows that this region performed slightly better in the later period. In Period A (2000–2005), the average Attractiveness to Foreign Investors score for the region is –0.7571 and it has increased to –0.6972 in Period B (2006–2012).

In Figure 3.10, Western China also lies in the third quadrant slightly above the 45-degree line, implying that this region performed somewhat better in the later period. In Period A (2000–2005), the average Attractiveness to Foreign Investors score for the region is –0.7553 and it has

Figure 3.10: ACI 2015 Competitiveness Over Time for Attractiveness to Foreign Investors

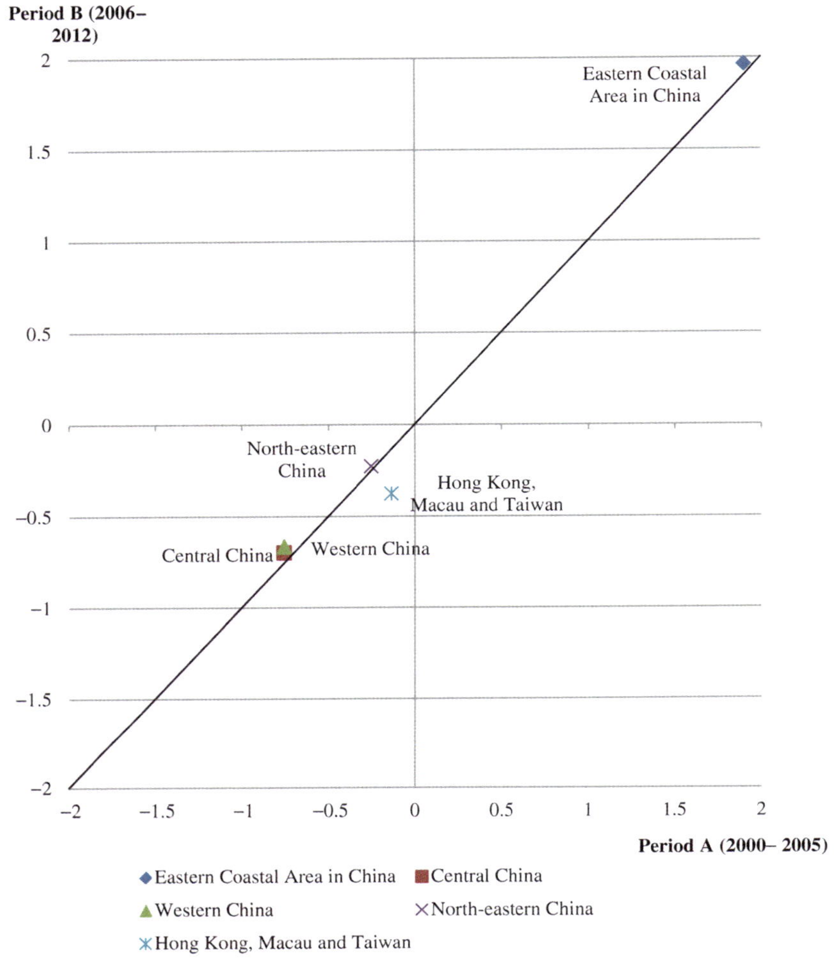

Source: Asia Competitiveness Institute

increased to –0.6645 in Period B (2006–2012). It is worth noting that the red square denoting Central China and the green triangle denoting Western China are almost overlapped in the third quadrant. The scores of these two regions are very close to each other, implying that there is a healthy competition between them.

As seen in Figure 3.10, North-eastern China is lying again in the third quadrant slightly above the 45-degree line, which indicates that this region performed better in the later period. In Period A (2000–2005), the average Attractiveness to Foreign Investors score for the region is –0.2529 and it has increased to –0.2261 in Period B (2006–2012).

Meanwhile, Hong Kong, Macau and Taiwan is below the 45-degree line. This shows that the region has deteriorated in terms of competitiveness over time as its average score declined from –0.1361 in Period A (2000–2005) to –0.3758 in Period B (2006–2012).

3.2.3 ACI 2015 Government and Institutional Setting Rankings and Scores

The Government and Institutional Setting ranking is presented in Table 3.5 with each region attaining the same position as in the Macroeconomic Stability ranking (see Table 3.3). The top performer of this environment is the Eastern Coastal Area with a score of 1.6060, while North-eastern China is ranked #5 with a score of –0.9775.

The average performing region is situated between the #2-ranked Hong Kong, Macau and Taiwan (0.7148) and the #3-ranked Western China (–0.5101), while the median is the #3-ranked Western China. The median score (–0.5101) is noticeably lower than the average score (0).

The distribution of scores in the Government and Institutional Setting ranking is uneven as well. This is especially distinct at the middle of the ranking. The difference in the scores between the #2-ranked Hong Kong, Macau and Taiwan and the #3-ranked Western China is 1.2249. Clearly, the Eastern Coastal Area as well as Hong Kong, Macau and Taiwan outperform other regions in Greater China. The scores of Western China, Central China and North-eastern China as a reflection of their competitiveness are relatively closer to each other.

Figure 3.11 shows the Government and Institutional Setting ranking of the five regions in Greater China from a geographical perspective. Regions are colour-coded according to their level of competitiveness, whereby more competitive regions are shaded darker while the less competitive regions have lighter colours. Since all regions in Greater China achieve the same respective rankings as in the Macroeconomic Stability environment, Figure 3.11 is essentially the same as Figure 3.6 except in a different colour scheme.

Table 3.6 presents the Government and Institutional Setting rankings and scores for the five regions in Greater China from 2000 to 2012. The Eastern Coastal Area has surpassed Hong Kong, Macau and Taiwan as #1 since 2001 with its score steadily increasing from 0.8756 in 2000 to 1.6060 in 2012. In contrast, the score of Hong Kong, Macau and Taiwan has been shrinking over the same period, from 1.4818 in 2000 to 0.7148 in 2012. Similar to the case of Overall Competitiveness, the gap in the scores between the #1-ranked and the #2-ranked region has widened, from a difference of 0.2214 in 2001 to 0.8912 in 2012.

North-eastern China has shown signs of diminishing competitiveness in Government and Institutional Setting as its ranking slips from #3 in the earliest four years to #4 from 2004 to

Table 3.5: ACI 2015 Government and Institutional Setting Ranking and Scores[20]

Rank	Region	Score
1	Eastern Coastal Area in China	1.6060
2	Hong Kong, Macau and Taiwan	0.7148
3	Western China	–0.5101
4	Central China	–0.8331
5	North-eastern China	–0.9775

Source: Asia Competitiveness Institute

[20]Result shown in Table 3.5 is based on 2012 data, collated and analysed in 2015.

Figure 3.11: ACI 2015 Government and Institutional Setting Ranking Map[21]

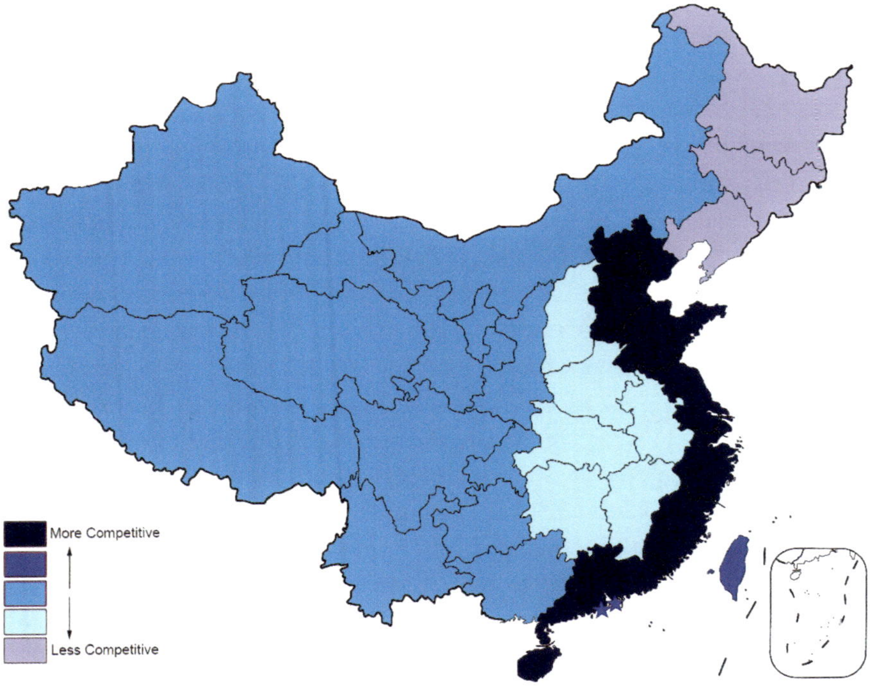

Source: Asia Competitiveness Institute edited from a map provided at http://bbs.eduol.cn/file.php?FileName=20128151347193336.jpg

2006 and then to the bottom position since 2007. At the same time, Western China has been catching up, overtaking North-eastern China to obtain the third position since 2004. Central China has also seen improvement in recent years, overtaking North-eastern China to obtain the fourth position since 2007.

Under the Government and Institutional Setting environment, there are two sub-environments, namely Government Policies and Fiscal Sustainability and Institutions, Governance and Leadership. Figure 3.12 allows us to compare the performances of the five regions in these two sub-environments. For the Government Policies and Fiscal Sustainability sub-environment, the Eastern Coastal Area performs exceptionally well with a high positive score of 1.9668 while the remaining regions have negative scores. With the lowest score in this sub-environment, Central China should work upon improving its Government Policies and Fiscal Sustainability. As for Institutions, Governance and Leadership, Hong Kong, Macau and Taiwan is at the top, followed by the Eastern Coastal Area.

For Government Policies and Fiscal Sustainability in Figure 3.13, the blue diamond dot representing the Eastern Coastal Area in China lies slightly above the 45-degree line. This means that the region performed somewhat better in the later period, Period B (2006–2012) as compared to Period A (2000–2005).

---

[21] Result shown in Figure 3.11 is based on 2012 data, collated and analysed in 2015.

Table 3.6: ACI 2015 Government and Institutional Setting Rankings and Scores: Evolution over Time[22]

	1		2		3	
Rank	**Region**	**Score**	**Region**	**Score**	**Region**	**Score**
Government and Institutional Setting						
2000	Hong Kong, Macau and Taiwan	1.4818	Eastern Coastal Area in China	0.8756	North-eastern China	–0.4722
2001	Eastern Coastal Area in China	1.3211	Hong Kong, Macau and Taiwan	1.0997	North-eastern China	–0.5604
2002	Eastern Coastal Area in China	1.3624	Hong Kong, Macau and Taiwan	1.0456	North-eastern China	–0.5230
2003	Eastern Coastal Area in China	1.4002	Hong Kong, Macau and Taiwan	1.0206	North-eastern China	–0.6393
2004	Eastern Coastal Area in China	1.3553	Hong Kong, Macau and Taiwan	1.0783	Western China	–0.6973
2005	Eastern Coastal Area in China	1.3334	Hong Kong, Macau and Taiwan	1.1029	Western China	–0.6947
2006	Eastern Coastal Area in China	1.3917	Hong Kong, Macau and Taiwan	1.0321	Western China	–0.6505
2007	Eastern Coastal Area in China	1.4570	Hong Kong, Macau and Taiwan	0.9494	Western China	–0.6353
2008	Eastern Coastal Area in China	1.5306	Hong Kong, Macau and Taiwan	0.8589	Western China	–0.7096
2009	Eastern Coastal Area in China	1.6113	Hong Kong, Macau and Taiwan	0.7349	Western China	–0.6464
2010	Eastern Coastal Area in China	1.6319	Hong Kong, Macau and Taiwan	0.6926	Western China	–0.5894
2011	Eastern Coastal Area in China	1.6053	Hong Kong, Macau and Taiwan	0.7295	Western China	–0.5595
2012	Eastern Coastal Area in China	1.6060	Hong Kong, Macau and Taiwan	0.7148	Western China	–0.5101

(Continued)

[22] Result shown in Table 3.6 is based on 2000–2012 data, collated and analysed in 2015.

Table 3.6: *(Continued)*

	4		5	
Rank	**Region**	**Score**	**Region**	**Score**
2000	Western China	−0.8059	Central China	−1.0793
2001	Western China	−0.9099	Central China	−0.9505
2002	Western China	−0.9347	Central China	−0.9504
2003	Western China	−0.8593	Central China	−0.9223
2004	North-eastern China	−0.8099	Central China	−0.9264
2005	North-eastern China	−0.8136	Central China	−0.9280
2006	North-eastern China	−0.8517	Central China	−0.9216
2007	Central China	−0.8400	North-eastern China	−0.9309
2008	Central China	−0.7909	North-eastern China	−0.8890
2009	Central China	−0.8247	North-eastern China	−0.8751
2010	Central China	−0.8204	North-eastern China	−0.9147
2011	Central China	−0.8572	North-eastern China	−0.9181
2012	Central China	−0.8331	North-eastern China	−0.9775

Source: Asia Competitiveness Institute

Figure 3.12: ACI 2015 Sub-environment Scores of Government and Institutional Setting[23]

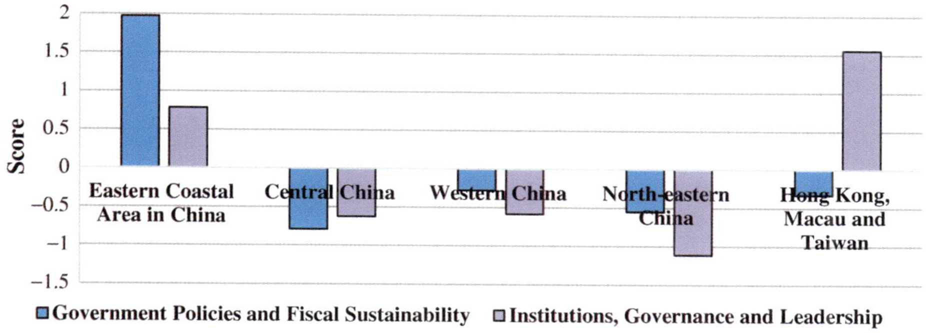

Source: Asia Competitiveness Institute

Figure 3.13: ACI 2015 Competitiveness Over Time for Government Policies and Fiscal Sustainability

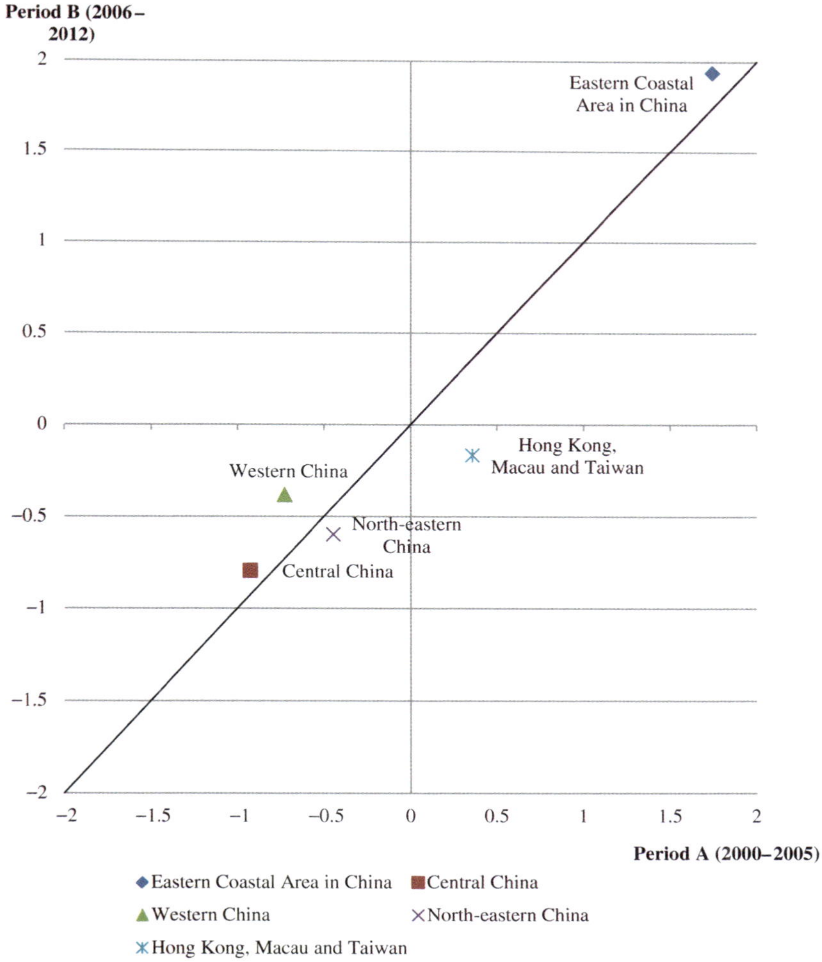

Source: Asia Competitiveness Institute

[23]Results shown in Figures 3.12–3.14 are based on 2012 data, collated and analysed in 2015.

The red square denoting Central China lies slightly above the 45-degree line, indicating that the region has improved on its competitiveness in Period B (2006–2012) relative to Period A (2000–2005). The green triangle denoting Western China lies above the 45-degree line as well, implying that the region has become more competitive over time.

As seen in Figure 3.13, the black cross denoting North-eastern China lies slightly below the 45-degree line, signifying that the region performed better in the earlier period, Period A (2000–2005).

The blue asterisk representing the Hong Kong, Macau and Taiwan lies far below the 45-degree line. This means that the performance of this region has deteriorated significantly over the two periods.

For Figure 3.14 on Institutions, Governance and Leadership, the blue diamond dot denoting the Eastern Coastal Area in China lies above the 45-degree line. However, as it is not at the top right corner of Figure 3.14 like the other sub-environments (Figure 3.5 and Figures 3.8–3.10),

Figure 3.14: ACI 2015 Competitiveness Over Time for Institutions, Governance and Leadership

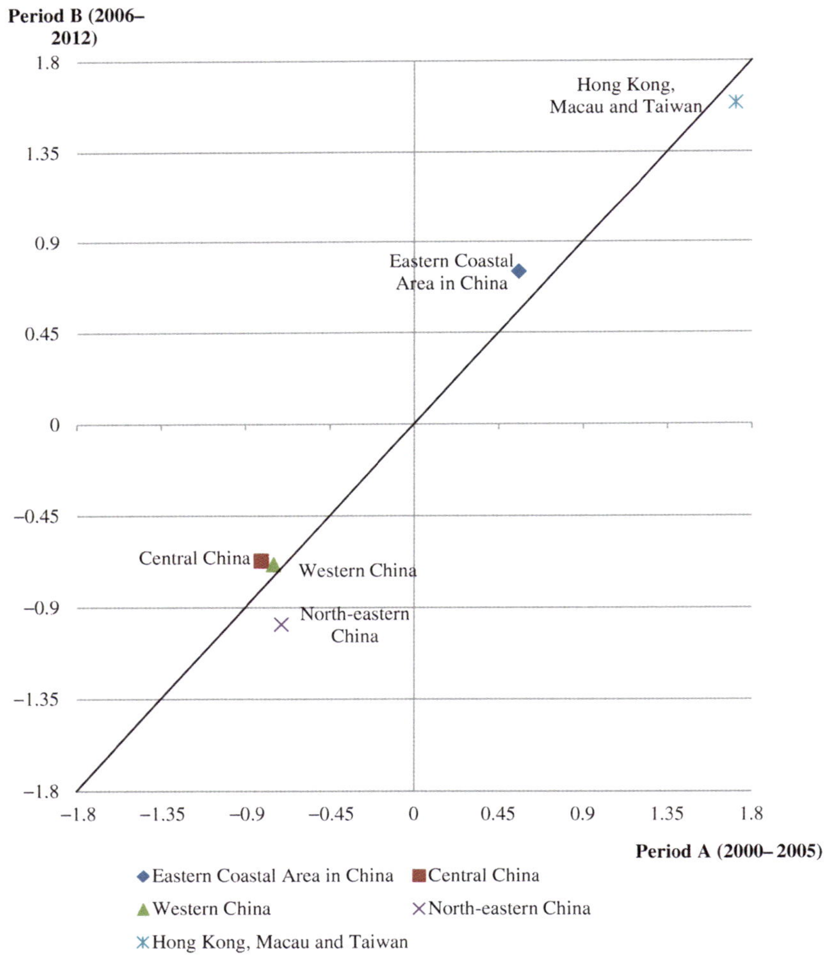

Source: Asia Competitiveness Institute

but sits near the centre of Figure 3.14, this means that although the region's competitiveness in this particular sub-environment has improved over time, it is still not as competitive as Hong Kong, Macau and Taiwan which lies at the top right corner of the Figure 3.14.

Figure 3.14 shows that Central China is slightly above the 45-degree line, implying that the region performs somewhat better in this sub-environment during Period B (2006–2012). The red square representing Central China is close to the cross for North-eastern China and the green triangle as Western China. This shows that their competitiveness levels, as indicated by their respective scores, are rather close, portending some rivalry as a healthy way to strengthen the overall competitiveness of China as a whole.

As seen in Figure 3.14, Western China is also a tad above the 45-degree line, implying that the region performs slightly better in this sub-environment in Period B (2006–2012). Meanwhile, North-eastern China is significantly below the 45-degree line, which indicates that the region performed considerably better in this sub-environment during Period A (2000–2005).

Meanwhile, the blue asterisk denoting Hong Kong, Macau and Taiwan lies slightly below the 45-degree line. Though Hong Kong, Macau and Taiwan is more competitive than the remaining regions in Greater China, as shown in its position on the top right corner of Figure 3.14, its competitiveness in this particular sub-environment has deteriorated slightly in Period B (2006–2012) as compared to Period A (2000–2005).

3.2.4 ACI 2015 Financial, Businesses and Manpower Conditions Rankings and Scores

Table 3.7 presents the Financial, Businesses and Manpower Conditions ranking for the five regions in Greater China. The top performer in this environment is Hong Kong, Macau and Taiwan, with a score of 1.3300. North-eastern China is ranked last at #5 with a score of –0.8573. This result is in line with the Financial, Businesses and Manpower Conditions ranking for the 34 Greater China economies where Hong Kong, Taiwan and Macau are ranked #1, #2 and #3, respectively.

The average performing region (with a standardised score of 0) is situated between the #2-ranked Eastern Coastal Area in China (1.1115) and the #3-ranked Western China (–0.7386), while the median score is –0.7386, which is the score of Western China. The median score for Financial, Businesses and Manpower Conditions ranking is hence lower than the average score (0).

Table 3.7: ACI 2015 Financial, Businesses and Manpower Conditions Ranking and Scores[24]

Rank	Region	Score
1	Hong Kong, Macau and Taiwan	1.3300
2	Eastern Coastal Area in China	1.1115
3	Western China	–0.7386
4	Central China	–0.8456
5	Northeastern China	–0.8573

Source: Asia Competitiveness Institute

[24]Result shown in Table 3.7 is based on 2012 data, collated and analysed in 2015.

The largest gap in the scores is that between the #2-ranked Eastern Coastal Area and #3-ranked Western China with a difference of 1.8501. This implies that Hong Kong, Macau and Taiwan and the Eastern Coastal Area are far ahead of the remaining three regions in Greater China. Interestingly, the scores of Central and North-eastern China are quite close to each other. This suggests that it is highly possible for North-eastern China to surpass Central China in the Financial, Businesses and Manpower Conditions ranking in the near future.

Figure 3.15 shows the Financial, Businesses and Manpower Conditions ranking of the five regions in Greater China from a geographical perspective. Regions are colour-coded according to their level of competitiveness, whereby more competitive regions are shaded darker while the less competitive regions have lighter colours. Unlike the rest of the environments, Hong Kong, Macau and Taiwan as one region is ranked #1 for Financial, Businesses and Manpower Conditions.

The rankings and scores of the five regions in Greater China for Financial, Businesses and Manpower Conditions from 2000 to 2012 are shown in Table 3.8. Three regions have retained their respective rankings throughout the 13-year period (2000–2012), namely Hong Kong, Macau and Taiwan, the Eastern Coastal Area and North-eastern China at #1, #2 and #5, respectively. Western and Central China have obtained the #3 and #4 places, respectively, for most of the years.

Figure 3.15: ACI 2015 Financial, Businesses and Manpower Conditions Ranking Map[25]

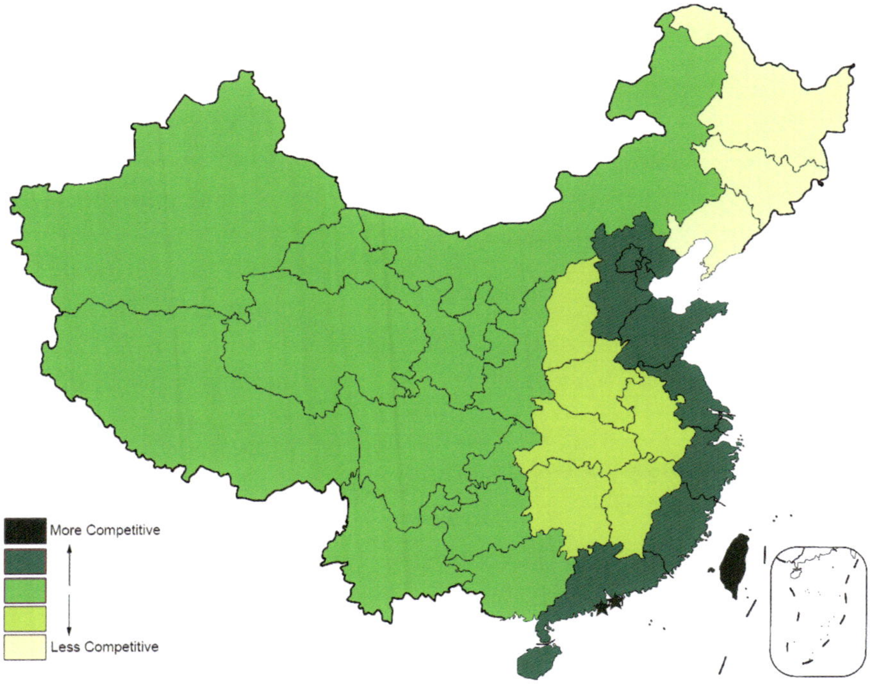

Source: Asia Competitiveness Institute edited from a map provided at http://bbs.eduol.cn/file.php?FileName=20128151347193336.jpg

[25]Result shown in Figure 3.15 is based on 2012 data, collated and analysed in 2015.

Table 3.8: ACI 2015 Financial, Businesses and Manpower Conditions Rankings and Scores: Evolution over Time[26]

	1		2		3	
Rank	**Region**	**Score**	**Region**	**Score**	**Region**	**Score**
Financial, Businesses and Manpower Conditions						
2000	Hong Kong, Macau and Taiwan	1.8463	Eastern Coastal Area in China	0.2756	Central China	–0.6413
2001	Hong Kong, Macau and Taiwan	1.8191	Eastern Coastal Area in China	0.3476	Western China	–0.6763
2002	Hong Kong, Macau and Taiwan	1.7543	Eastern Coastal Area in China	0.3969	Central China	–0.3524
2003	Hong Kong, Macau and Taiwan	1.7187	Eastern Coastal Area in China	0.5524	Western China	–0.6648
2004	Hong Kong, Macau and Taiwan	1.6065	Eastern Coastal Area in China	0.7455	Western China	–0.7055
2005	Hong Kong, Macau and Taiwan	1.5661	Eastern Coastal Area in China	0.8083	Western China	–0.7231
2006	Hong Kong, Macau and Taiwan	1.6130	Eastern Coastal Area in China	0.7359	Western China	–0.7059
2007	Hong Kong, Macau and Taiwan	1.5801	Eastern Coastal Area in China	0.7892	Western China	–0.7235
2008	Hong Kong, Macau and Taiwan	1.4972	Eastern Coastal Area in China	0.9087	Western China	–0.7674
2009	Hong Kong, Macau and Taiwan	1.4375	Eastern Coastal Area in China	0.9860	Western China	–0.7633
2010	Hong Kong, Macau and Taiwan	1.4076	Eastern Coastal Area in China	1.0230	Western China	–0.7792
2011	Hong Kong, Macau and Taiwan	1.3515	Eastern Coastal Area in China	1.0887	Western China	–0.7644
2012	Hong Kong, Macau and Taiwan	1.3300	Eastern Coastal Area in China	1.1115	Western China	–0.7386

(*Continued*)

[26]Result shown in Table 3.8 is based on 2000–2012 data, collated and analysed in 2015.

Table 3.8: (Continued)

	4		5	
Rank	**Region**	**Score**	**Region**	**Score**
2000	Western China	−0.6784	North-eastern China	−0.8022
2001	Central China	−0.7133	North-eastern China	−0.7771
2002	Western China	−0.7923	North-eastern China	−1.0065
2003	Central China	−0.7362	North-eastern China	−0.8701
2004	Central China	−0.7516	North-eastern China	−0.8949
2005	Central China	−0.7638	North-eastern China	−0.8875
2006	Central China	−0.7548	North-eastern China	−0.8881
2007	Central China	−0.7863	North-eastern China	−0.8595
2008	Central China	−0.7953	North-eastern China	−0.8432
2009	Central China	−0.8133	North-eastern China	−0.8468
2010	Central China	−0.7987	North-eastern China	−0.8526
2011	Central China	−0.8344	North-eastern China	−0.8413
2012	Central China	−0.8456	North-eastern China	−0.8573

Source: Asia Competitiveness Institute

Figure 3.16: ACI 2015 Sub-environment Scores of Financial, Businesses and Manpower Conditions[27]

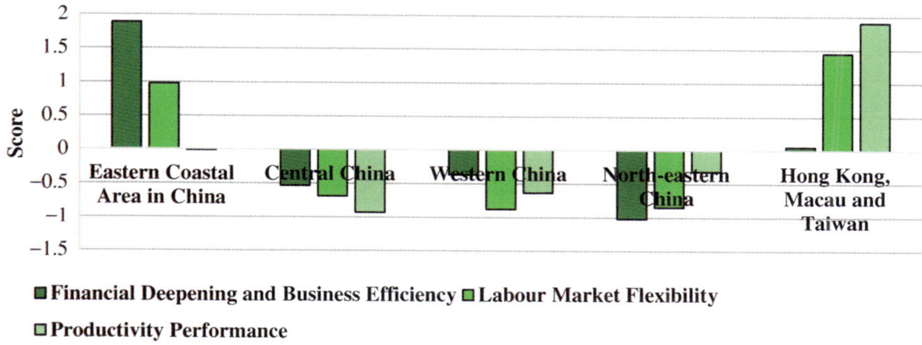

Source: Asia Competitiveness Institute

Though the rankings of Hong Kong, Macau and Taiwan and the Eastern Coastal Area have maintained the same over the years, their gap in competitiveness has shrunk. The Eastern Coastal Area has been steadily catching up, closing the gap between them. In 2000, the difference in the scores of the two regions is 1.5707 and it has diminished to 0.2185 by 2012. Based on the trend shown, it is likely that in the foreseeable future, the Eastern Coastal Area will surpass Hong Kong, Macau and Taiwan in terms of Financial, Businesses and Manpower Conditions.

There are three sub-environments under the Financial, Businesses and Manpower Conditions, namely Financial Deepening and Business Efficiency; Labour Market Flexibility; and Productivity Performance. Both the Eastern Coastal Area in China and Hong Kong, Macau and Taiwan perform well in Labour Market Flexibility (see Figure 3.16). However, Hong Kong, Macau and Taiwan should further improve its Financial Deepening and Business Efficiency, while the Eastern Coastal Area should work upon its Productivity Performance in order to improve their scores for the Financial, Businesses and Manpower Conditions environment. The scores of Central, Western and North-eastern China in these three sub-environments are all negative, implying that they have below average performance in this environment.

As shown in Figure 3.17, for Financial Deepening and Business Efficiency, the blue diamond dot denoting the Eastern Coastal Area in China lies above the 45-degree line, implying that the region performed better in Period B (2006–2012). The average score for the region in Period A (2000–2005) is 1.2154 relative to 1.7350 in Period B (2006–2012).

Meanwhile, the red square denoting Central China is slightly below the 45-degree line, suggesting that the competitiveness of this region has decreased slightly over time. During Period A (2000–2005), the average score of Central China for this sub-environment is –0.5369, and it has dropped to –0.6184 in Period B (2006–2012).

In Figure 3.17, the green triangle denoting Western China is above the 45-degree line, which indicates that this region has become more competitive over time. Indeed, the average score of Western China for this sub-environment increased from –0.7279 in Period A (2000–2005) to –0.4994 in Period B (2006–2012).

[27]Results shown in Figures 3.16–3.19 are based on 2012 data, collated and analysed in 2015.

Figure 3.17: ACI 2015 Competitiveness Over Time for Financial Deepening and Business Efficiency

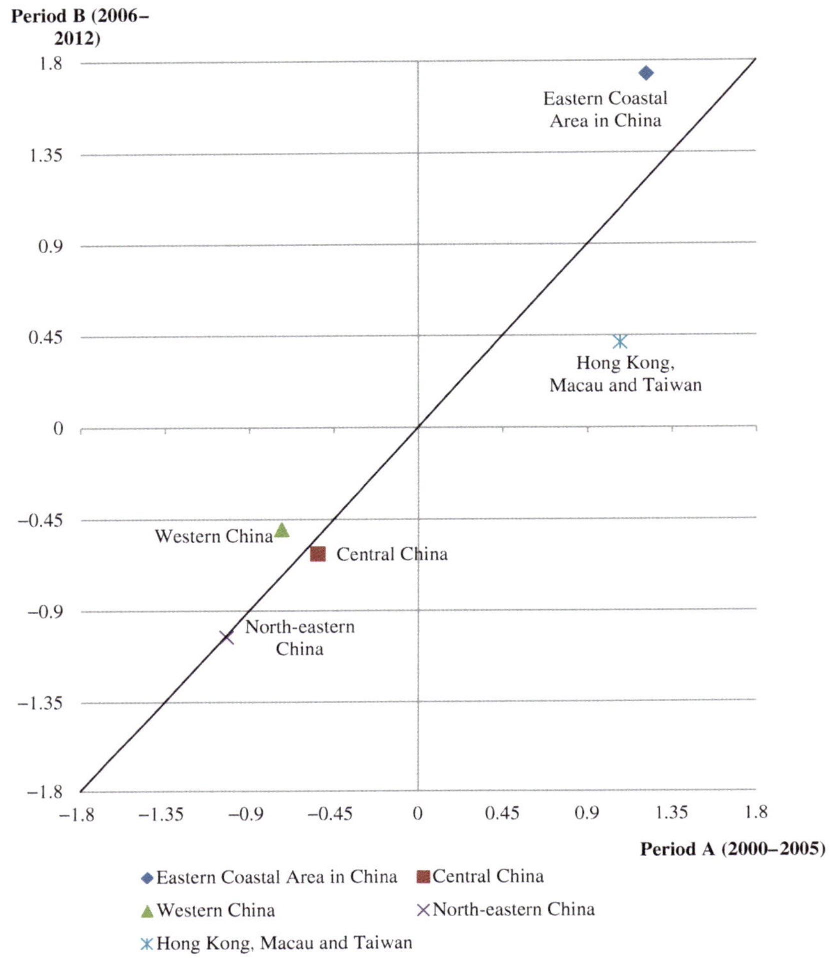

Source: Asia Competitiveness Institute

The black cross denoting North-eastern China falls on the bottom left corner of Figure 3.17, almost exactly on the 45-degree line. This shows that the average scores of the region remained roughly the same for the two periods (2000–2005 and 2006–2012). The region is not only noticeably less competitive than other regions in this particular sub-environment (as shown in its position on the bottom left corner of Figure 3.17, far away from the rest of the regions), but also has shown no improvement over time.

As seen in Figure 3.17, the blue asterisk denoting Hong Kong, Macau and Taiwan lies far below the 45-degree line, implying that its competitiveness has deteriorated considerably. The average score for the region declined from 1.0731 in Period A (2000–2005) to 0.4129 in Period B (2006–2012).

Figure 3.18 shows the Competitiveness Over Time for Labour Market Flexibility. In this sub-environment, the competitiveness of the Eastern Coastal Area in China has improved in Period B (2006–2012) as it lies above the 45-degree line. However, lying near the centre of Figure 3.18,

Figure 3.18: ACI 2015 Competitiveness Over Time for Labour Market Flexibility

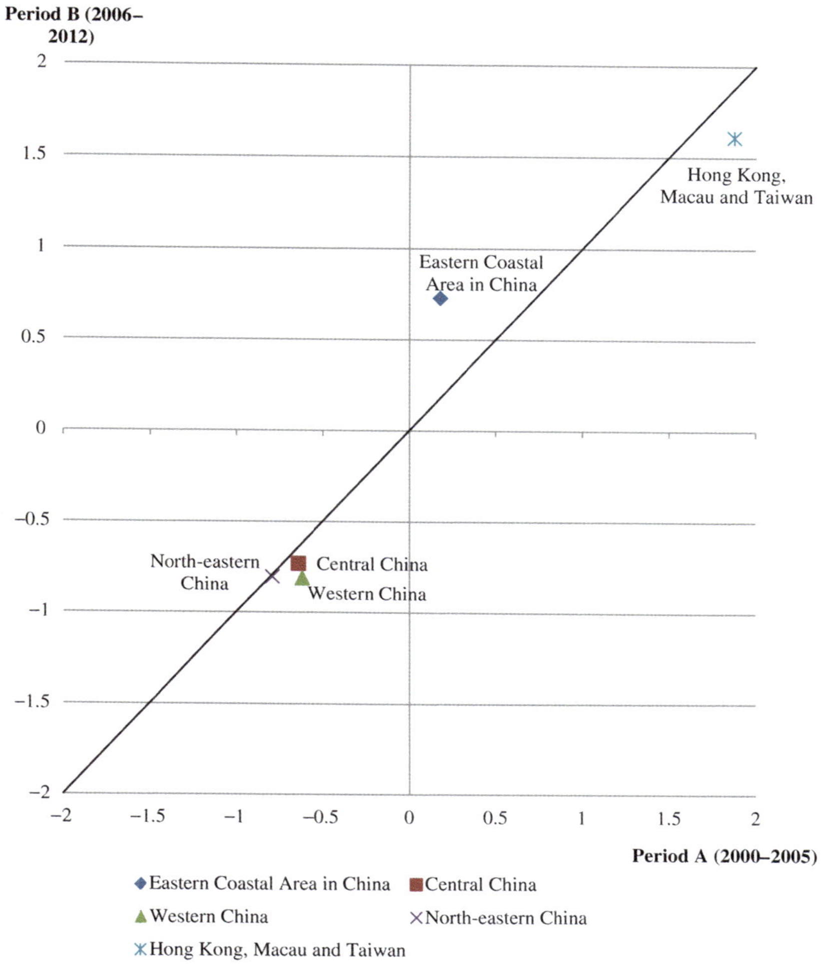

Source: Asia Competitiveness Institute

in contrast to Hong Kong, Macau and Taiwan on the top right corner of Figure 3.18, the Eastern Coastal Area in China is still not as competitive as Hong Kong, Macau and Taiwan combined.

In Figure 3.18, Central China lies below the 45-degree line, signifying that the competitiveness level of the region has deteriorated in Period B (2006–2012) relative to Period A (2000–2005). It is worth noting that Central China, Western China, and Northeastern China are clustered together in the third quadrant. The scores of these three regions are close to each other, a sign of healthy competition within China to be more competitive *vis-à-vis* the rest of the world.

As seen in Figure 3.18, the competitiveness level of Western China in Period B (2006–2012) is lower, as the region lies below the 45-degree line. Meanwhile, North-eastern China has remained stagnant in terms of competitiveness over time, as shown in its position on the 45-degree line.

In this sub-environment, Hong Kong, Macau and Taiwan lies below the 45-degree line, implying that the competitiveness of Hong Kong, Macau and Taiwan has weakened in Period B

(2006–2012) relative to Period A (2000–2005). However, it is still the most competitive region in Greater China, as indicated by its position on the top right corner of Figure 3.18.

For Productivity Performance, Figure 3.19 shows that the blue diamond dot representing the Eastern Coastal Area in China lies on the 45-degree line and is rather close to the centre of Figure 3.19. This implies that the region is not as competitive as Hong Kong, Macau and Taiwan (at the top right corner of Figure 3.19) and its competitiveness has remained stagnant over time.

Figure 3.19 shows that the red square denoting Central China is a tad below the 45-degree line, implying that the region performed slightly better in Period A (2000–2005). The average score of the region for the earlier period is –0.6852 while for the later period (2006–2012) it is –0.8079.

As seen in Figure 3.19, the green triangle denoting Western China is also slightly below the 45-degree line, which indicates that the region was somewhat more competitive in Period A

Figure 3.19: ACI 2015 Competitiveness Over Time for Productivity Performance

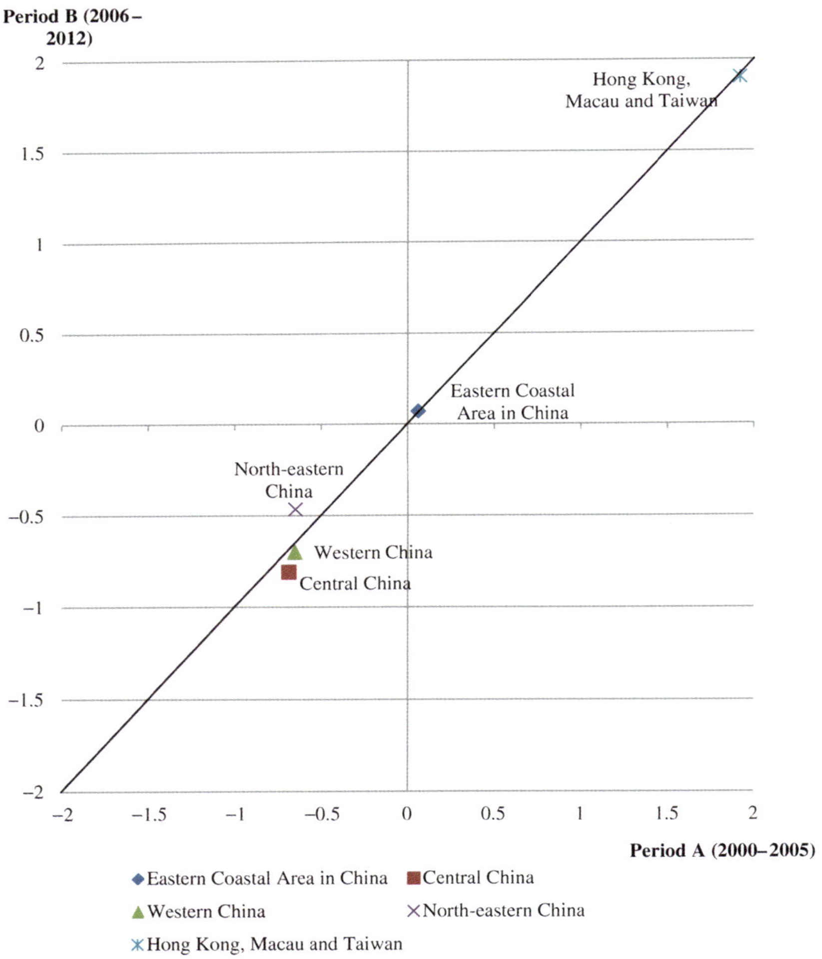

Source: Asia Competitiveness Institute

(2000–2005) as compared to Period B (2006–2012). Indeed, the average score of the region declined from –0.6490 in Period A (2000–2005) to –0.6997 in Period B (2006–2012).

Meanwhile, the black cross denoting North-eastern China is above the 45-degree line, signifying that the region performed better in Period B (2006–2012) relative to Period A (2000–2005). The average score of the region rose from –0.6475 in the earlier period to –0.4650 in the later period.

The blue asterisk representing Hong Kong, Macau and Taiwan lies on the 45-degree line in Figure 3.19 and at the top right corner. This shows the region not only has performed well in both periods but is also the most competitive region in Greater China.

3.2.5 ACI 2015 Quality of Life and Infrastructure Development Rankings and Scores

Table 3.9 shows the ranking of Quality of Life and Infrastructure Development. The Eastern Coastal Area in China is ranked first with the highest score (1.8419) among the five regions in Greater China while Central China is ranked last with a score of –1.1470. Although North-eastern China is ranked #3 in this environment, its score of –0.0911 is not far from the score of the #2-ranked Hong Kong, Macau and Taiwan (–0.0768).

The average performing region (with a standardised score of 0) is somewhere between the #1-ranked Eastern Coastal Area in China (1.8419) and the #2-ranked Hong Kong, Macau and Taiwan (–0.0768). The median score for Quality of Life and Infrastructure Development ranking is –0.0911. For this environment, the median score (–0.0911) is only slightly lower than the average score (0).

The largest gap in the scores is that between the #1- and the #2-ranked region as the scores of the Eastern Coastal Area and Hong Kong, Macau and Taiwan differ by 1.9187. This shows that the Eastern Coastal Area is far more competitive than the other regions in Greater China. Another large gap in the scores is between the #4- and the #5-ranked region as the score of Western China is higher than Central China by 0.6200.

Figure 3.20 provides a geographical snapshot of the Quality of Life and Infrastructure Development ranking. Regions that are more competitive are in darker colours, while less competitive regions have lighter shades.

Table 3.9: ACI 2015 Quality of Life and Infrastructure Development Ranking and Scores[28]

Rank	Region	Score
1	Eastern Coastal Area in China	1.8419
2	Hong Kong, Macau and Taiwan	–0.0768
3	North-eastern China	–0.0911
4	Western China	–0.5270
5	Central China	–1.1470

Source: Asia Competitiveness Institute

[28]Result shown in Table 3.9 is based on 2012 data, collated and analysed in 2015.

Figure 3.20: ACI 2015 Quality of Life and Infrastructure Development Ranking Map[29]

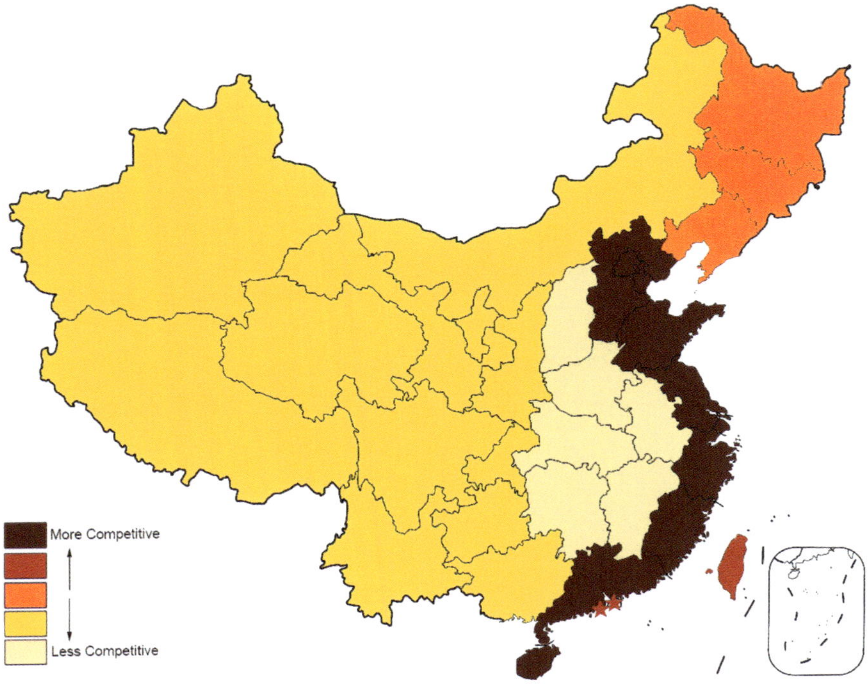

Source: Asia Competitiveness Institute edited from a map provided at http://bbs.eduol.cn/file.php?FileName=20128151347193336.jpg

Table 3.10 presents the Quality of Life and Infrastructure Development rankings and scores for the five regions of Greater China from 2000 to 2012. It enables us to track the competitiveness landscape within Greater China over the past decade. As seen in Table 3.10, the Eastern Coastal Area in China has been gradually improving its scores and has surpassed Hong Kong, Macau and Taiwan since 2004. North-eastern China, Western China and Central China have consistently maintained their third, fourth and bottom positions, respectively, throughout the 13 years.

Meanwhile, the ranking of Hong Kong, Macau and Taiwan has fallen from #1 in the first four years to #2 since 2004, accompanied by a drop in scores from 1.2871 in 2000 to –0.0768 in 2012. Hong Kong, Macau and Taiwan has changed from the top performer in Greater China to a below average performer, as indicated in its negative score in 2012. Although North-eastern China, Western China and Central China have maintained their rankings from 2000 to 2012, their competitiveness as indicated by their respective scores have risen over the years.

[29]Result shown in Figure 3.20 is based on 2012 data, collated and analysed in 2015.

Table 3.10: ACI 2015 Quality of Life and Infrastructure Development Rankings and Scores: Evolution over Time[30]

	1		2		3	
Rank	**Region**	**Score**	**Region**	**Score**	**Region**	**Score**
Quality of Life and Infrastructure Development						
2000	Hong Kong, Macau and Taiwan	1.2871	Eastern Coastal Area in China	1.0153	North-eastern China	−0.2233
2001	Hong Kong, Macau and Taiwan	1.3647	Eastern Coastal Area in China	0.9799	North-eastern China	−0.3504
2002	Hong Kong, Macau and Taiwan	1.1970	Eastern Coastal Area in China	1.1062	North-eastern China	−0.2359
2003	Hong Kong, Macau and Taiwan	1.2383	Eastern Coastal Area in China	1.0872	North-eastern China	−0.2904
2004	Eastern Coastal Area in China	1.1835	Hong Kong, Macau and Taiwan	1.1159	North-eastern China	−0.2390
2005	Eastern Coastal Area in China	1.4507	Hong Kong, Macau and Taiwan	0.8034	North-eastern China	−0.2064
2006	Eastern Coastal Area in China	1.4979	Hong Kong, Macau and Taiwan	0.7523	North-eastern China	−0.2147
2007	Eastern Coastal Area in China	1.5019	Hong Kong, Macau and Taiwan	0.8013	North-eastern China	−0.3692
2008	Eastern Coastal Area in China	1.6803	Hong Kong, Macau and Taiwan	0.5102	North-eastern China	−0.2990
2009	Eastern Coastal Area in China	1.6997	Hong Kong, Macau and Taiwan	0.3007	North-eastern China	−0.0063
2010	Eastern Coastal Area in China	1.8406	Hong Kong, Macau and Taiwan	0.0530	North-eastern China	−0.1635
2011	Eastern Coastal Area in China	1.8085	Hong Kong, Macau and Taiwan	0.1263	North-eastern China	−0.1897
2012	Eastern Coastal Area in China	1.8419	Hong Kong, Macau and Taiwan	−0.0768	North-eastern China	−0.0911

(Continued)

[30] Result shown in Table 3.10 is based on 2000–2012 data, collated and analysed in 2015.

Table 3.10: (*Continued*)

	4		5	
Rank	**Region**	**Score**	**Region**	**Score**
2000	Western China	–0.8144	Central China	–1.2647
2001	Western China	–0.8152	Central China	–1.1790
2002	Western China	–0.7588	Central China	–1.3085
2003	Western China	–0.7630	Central China	–1.2721
2004	Western China	–0.7347	Central China	–1.3256
2005	Western China	–0.7883	Central China	–1.2594
2006	Western China	–0.8270	Central China	–1.2085
2007	Western China	–0.7481	Central China	–1.1859
2008	Western China	–0.8082	Central China	–1.0834
2009	Western China	–0.8691	Central China	–1.1249
2010	Western China	–0.6581	Central China	–1.0721
2011	Western China	–0.5944	Central China	–1.1507
2012	Western China	–0.5270	Central China	–1.1470

Source: Asia Competitiveness Institute

Figure 3.21: ACI 2015 Sub-environment Scores of Quality of Life and Infrastructure Development[31]

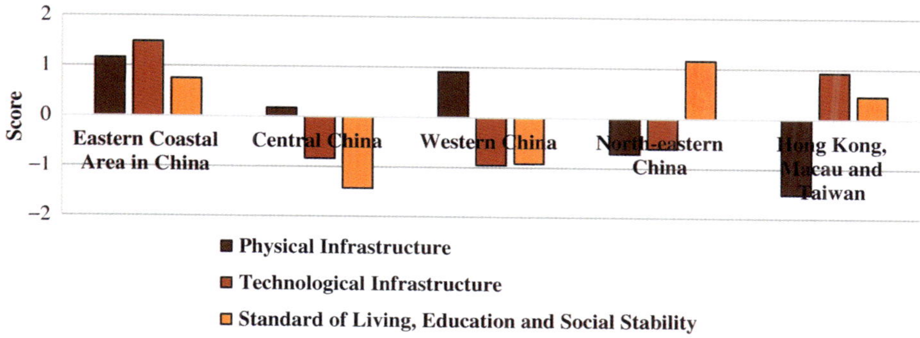

Source: Asia Competitiveness Institute

Quality of Life and Infrastructure Development has three sub-environments, namely Physical Infrastructure, Technological Infrastructure, and Standard of Living, Education and Social Stability. The outstanding performance of the Eastern Coastal Area in this environment can be explained by the region's strong performance in all of the three sub-environments, as indicated by their high scores (see Figure 3.21). North-eastern China possesses high score in Standard of Living, Education and Social Stability, which is the maximum for all regions in Greater China. It is worth mentioning that Western China performs relatively well in Physical Infrastructure, as it has the second highest score in Greater China after the Eastern Coastal Area.

As for Hong Kong, Macau and Taiwan, its less competitive Physical Infrastructure has dragged its score for Quality of Life and Infrastructure Development down. Central China should consider improving its Standard of Living, Education and Social Stability, as it has the lowest score for this sub-environment among the Greater China regions.

For Physical Infrastructure, the blue diamond dot denoting the Eastern Coastal Area in China lies slightly below the 45-degree line in Figure 3.22, which indicates that the region performed better in Period A (2000–2005) as compared to the Period B (2006–2012), with respective average scores of 1.2760 and 1.2142.

Central China, as denoted by the red square in Figure 3.22, lies in the first quadrant, on the 45-degree line. This implies that the region has not improved in terms of competitiveness over time. The two regions that are more competitive than Central China are the Eastern Coastal Area in China and Western China, whose respective dots are on the top right corner of Figure 3.22.

As seen in Figure 3.22, the green triangle denoting Western China lies in the first quadrant, slightly above the 45-degree line. This shows that the region performed somewhat better in Period B (2006–2012) relative to Period A (2000–2005). The only region that is more competitive than Western China is the Eastern Coastal Area, whose dot is at the top right corner of Figure 3.22.

[31] Results shown in Figures 3.21–3.24 are based on 2012 data, collated and analysed in 2015.

Figure 3.22: ACI 2015 Competitiveness Over Time for Physical Infrastructure

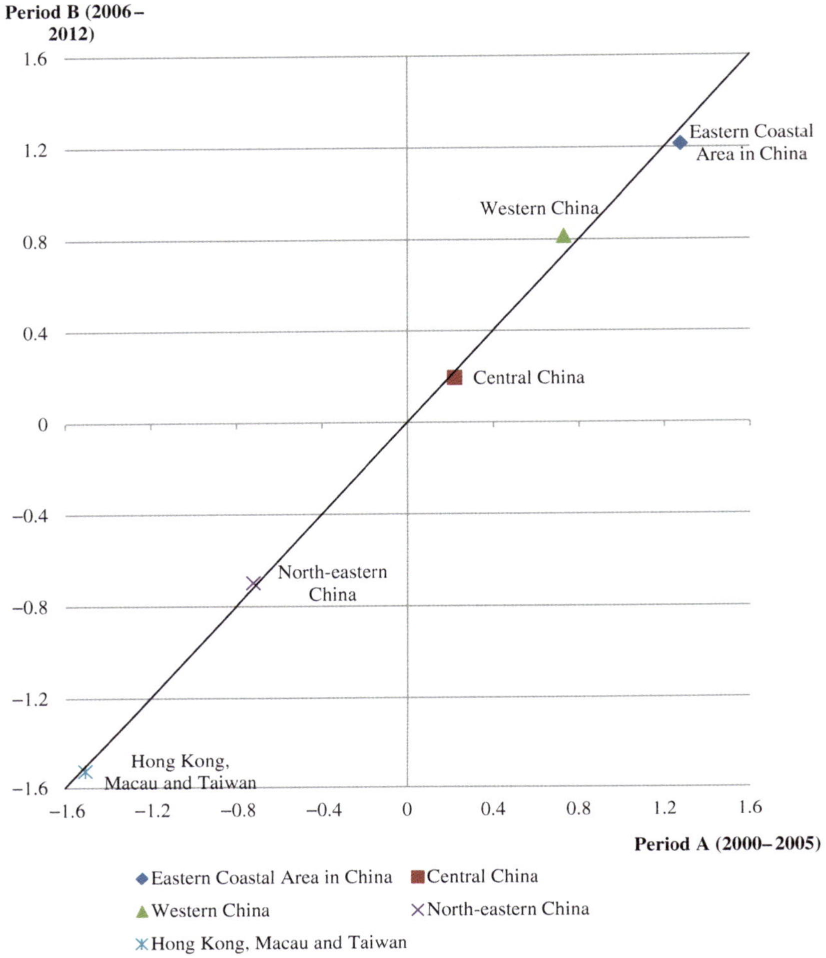

Source: Asia Competitiveness Institute

Figure 3.22 also shows that the competitiveness of North-eastern China has improved marginally in Period B (2006–2012). The average Physical Infrastructure score for the region increased from −0.7207 in Period A (2000–2005) to −0.6985 in Period B (2006–2012).

The blue asterisk denoting Hong Kong, Macau and Taiwan lies on the 45-degree line in Figure 3.22, which implies that the region's competitiveness has remained stagnant over time. However, unlike the other sub-environments, the blue asterisk lies at the bottom left corner of Figure 3.22, indicating that Hong Kong, Macau and Taiwan is the least competitive among all regions in Greater China in terms of Physical Infrastructure. This should serve as a reminder for the region to work upon its infrastructure.

Figure 3.23 shows the Competitiveness Over Time for Technological Infrastructure. The Eastern Coastal Area in China lies above the 45-degree line, implying that it has shown improvement in competitiveness in Period B (2006–2012) as compared to Period A (2000–2005). It is

Figure 3.23: ACI 2015 Competitiveness Over Time for Technological Infrastructure

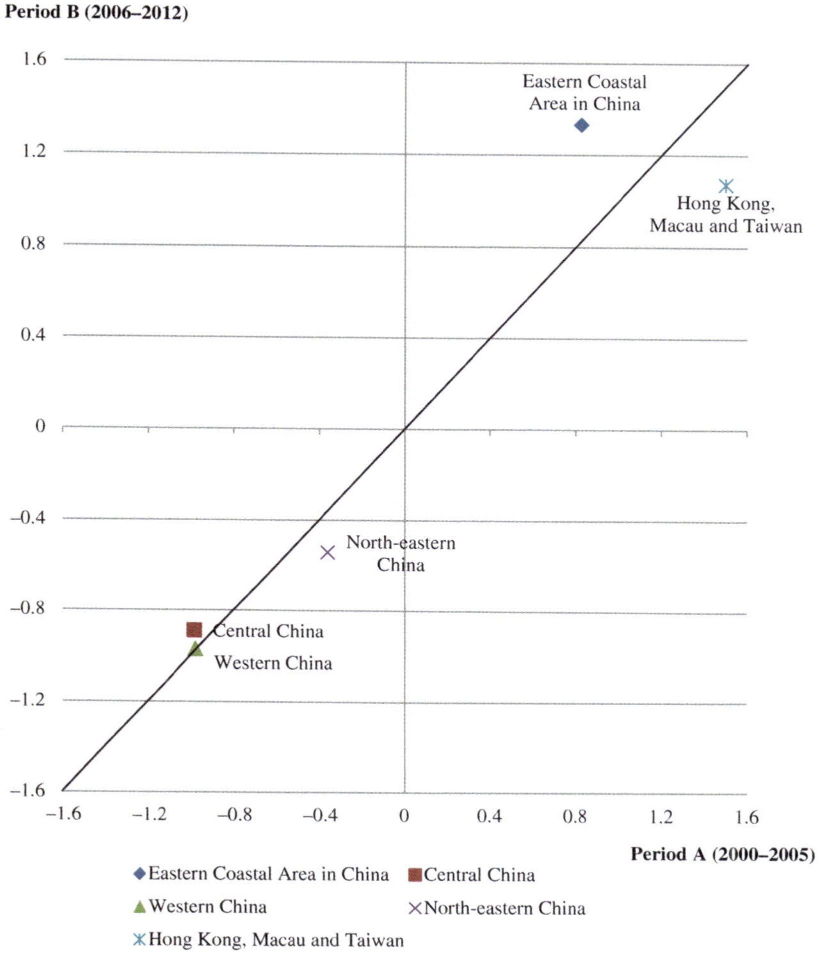

Source: Asia Competitiveness Institute

rather interesting that Hong Kong, Macau and Taiwan as one region is within the same quadrant as the Eastern Coastal Area, but below the 45-degree line. Although the average score of Hong Kong, Macau and Taiwan is higher than that of the Eastern Coastal Area in Period A (2000–2005), the Eastern Coastal Area has since surpassed Hong Kong, Macau and Taiwan in Period B (2006–2012).

As seen in Figure 3.23, Central China lies slightly above the 45-degree line, which indicates that the region performed better in Period B (2006–2012) than Period A (2000–2005). The average score of the region for the earlier period is –0.9841 while for the later period is –0.8905.

Meanwhile, Western China lies almost exactly on the 45-degree line. This means that the region has remained relatively stagnant in terms of competitiveness over time. The average score of the region for the earlier period is –0.9791 while that for the later period is –0.9722. The black cross denoting North-eastern China lies below the 45-degree line, which implies that the region was more competitive in Period A (2000–2005) relative to Period B (2006–2012).

Hong Kong, Macau and Taiwan lies far below the 45-degree line, indicating that its competitiveness has deteriorated over time. The Eastern Coastal Area is within the same quadrant as Hong Kong, Macau and Taiwan, but above the 45-degree line. As noted earlier, Hong Kong, Macau and Taiwan has been overtaken in this sub-environment by the Eastern Coastal Area in China in Period B (2006–2012).

Figure 3.24 shows the Competitiveness Over Time for Standard of Living, Education and Social Stability. Unlike the rest of the sub-environments, the blue diamond dot denoting the Eastern Coastal Area lies in the second quadrant, way above the 45-degree line. This shows that the region has improved tremendously in Period B (2006–2012) as seen in the large vertical distance between the blue diamond dot and the 45-degree line. However, the region is still less

Figure 3.24: ACI 2015 Competitiveness Over Time for Standard of Living, Education and Social Stability

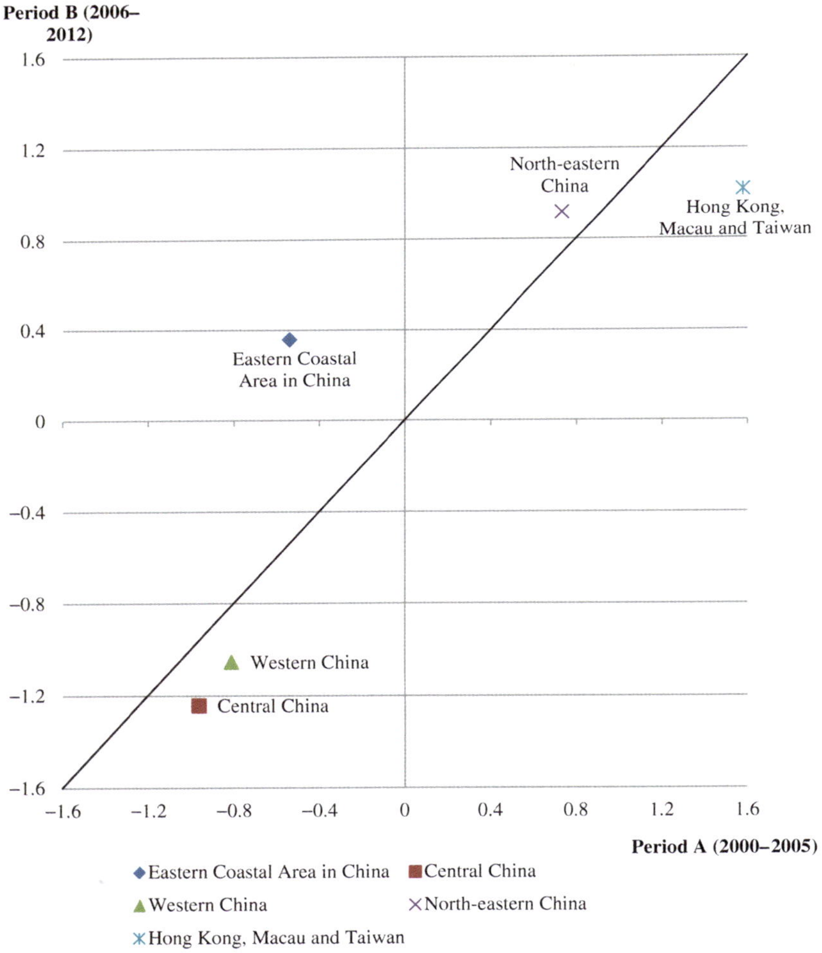

Source: Asia Competitiveness Institute

competitive in terms of Standard of Living, Education and Social Stability as compared to North-eastern China which lies at the top right corner of Figure 3.24.

Meanwhile, the red square representing Central China is considerably below the 45-degree line, indicating that the region's competitiveness has deteriorated significantly over time, as seen in the relatively large vertical distance between the red square and the 45-degree line.

Similar to the case of Central China, in Figure 3.24, the green triangle representing Western China is also considerably below the 45-degree line. This shows that the region has also deteriorated in terms of competitiveness over time, as seen in the relatively large vertical distance between the green triangle and the 45-degree line.

Unlike the rest of the sub-environments, the black cross denoting North-eastern China lies in the first quadrant in Figure 3.24, above the 45-degree line. The region, which is located on the top right corner of Figure 3.24, is much more competitive than the rest of the regions in mainland China. The performance of North-eastern China in this sub-environment has also improved over the two periods.

In Figure 3.24, the blue asterisk denoting Hong Kong, Macau and Taiwan lies in the first quadrant, way below the 45-degree line. This implies that the region has deteriorated tremendously in terms of competitiveness in Period B (2006–2012) as seen in the large vertical distance between the blue asterisk and the 45-degree line. However, Hong Kong, Macau and Taiwan remains the most competitive region in this sub-environment in both periods, although the gap in the average scores between the region and North-eastern China has shrunk significantly, from 0.8474 in Period A (2000–2005) to 0.0981 in Period B (2006–2012).

3.3 CONCLUDING REMARKS

In summary, we have noted three interesting observations. Firstly, the Eastern Coastal Area has surpassed Hong Kong, Macau and Taiwan as #1 since 2002 with their gaps in competitiveness continue to widen over the years. Secondly, it is heartening to see the progressive improvement in the competitiveness of Western China as it has replaced North-eastern China as the #3-ranked region in 2007 and since 2010. Lastly, Central China has consistently ranked last among the five regions in Greater China with only marginal improvement in scores.

In order to find out the common and distinct strengths and weaknesses of each region, we have categorised the five regions in Greater China into three groups according to the 2015 Overall Competitiveness ranking, namely the top-performing group, which comprises the Eastern Coastal Area in China and Hong Kong, Macau and Taiwan; the middle-performing group, which are Western and North-eastern China; and the bottom-performing region, namely Central China. Common to both top-performing regions, they excel in Openness to Trade and Services; Institutions, Governance and Leadership; Labour Market Flexibility; Technological Infrastructure; and Standard of Living, Education and Social Stability. The shortcomings of Hong Kong, Macau and Taiwan that made less competitive than the Eastern Coastal Area is its poor Regional Economic Vibrancy; Attractiveness to Foreign Investors; Government Policies and Fiscal Sustainability; as well as Physical Infrastructure. That said, the Eastern Coastal Area should work upon improving its Productivity Performance.

Each middle-performing region has their unique strengths and weaknesses. Western China has outstanding Physical Infrastructure while North-eastern China enjoys remarkable Standard of Living, Education and Social Stability. Nevertheless, both regions have ample room for improvement in terms of Openness to Trade and Services, Institutions, Governance and Leadership, Labour Market Flexibility, and Technological Infrastructure.

All the sub-environment scores of the bottom-performing region, Central China, are negative, except for Regional Economic Vibrancy and Physical Infrastructure which are slightly positive. This shows that the Regional Economic Vibrancy and Physical Infrastructure of Central China have performed above average. Some areas for improvement for the region include Government Policies and Fiscal Sustainability, Productivity Performance, and Standard of Living, Education and Social Stability. The challenging situation is exacerbated by the net outflow of skilled labour, highly educated talents and entrepreneurs from Central China usually to the more developed Eastern Coastal Area in China.

REFERENCES

Chan, Chak K. and Xiaohong Yao. 2008. "Air Pollution in Mega Cities in China." *Atmospheric Environment* 42(1): 1–42.

Fan, C. Cindy. 1997. "Uneven Development and Beyond: Regional Development Theory in Post-Mao China." *International Journal of Urban and Regional Research* 21(4): 620–639.

Fan, C. Cindy and Mingjie Sun. 2008. "Regional Inequality in China, 1978–2006." *Eurasian Geography and Economics* 49(1): 1–18.

Fang, Cai, Dewen Wang and Qu Yue. 2009. "Flying Geese within Borders: How China Sustains Its Labour-Intensive Industries." *Economic Research Journal* 9.

Fujita, Masahisa and Dapeng Hu. 2001. "Regional Disparity in China 1985–1994: The Effects of Globalization and Economic Liberalization." *The Annals of Regional Science* 35(1): 3–37.

Guo, Ji-fu, Ying Liu and Liu Yu. "Traffic Congestion in Large Metropolitan Area in China." *Urban Transport of China* 2: 8–14.

Ha, Wei, Junjian Yi and Junsen Zhang. "Brain Drain, Brain Gain, and Economic Growth in China." *China Economic Review* 37(February): In Press.

Kojima, Kiyoshi. 2000. "The 'Flying Geese' Model of Asian Economic Development: Origin, Theoretical Extensions, and Regional Policy Implications." *Journal of Asian Economics* 11(4): 375–401.

Li, Shantong and Zhaoyuan Xu. 2008. "The Trend of Regional Income Disparity in the People's Republic of China." *Asian Development Bank Institute Discussion Paper* 85. Tokyo.

Li, Yining. 2014. "社会和谐红利是最大的制度红利 Social Harmony Is the Biggest Dividend from System (in Chinese)." *《现代国企研究》 Modern State-Owned Enterprise Research* 4.

Li, Yining, Justin Yifu Lin, Yongnian Zheng, Yansheng Zhang, Min Tang, Monan Zhang, Zhicheng Bao, Jincheng Xiao, Xianjie Shen and Zhiyuan Sun. 2015. *Understand One Belt, One Road. Understand One Belt, One Road*. Beijing: CITIC Publishing Group.

Lin, Justin Yifu. 2011. "China and the Global Economy." *China Economic Journal* 4(1): 1–14.

OECD. 2013. "Trust in Government, Policy Effectiveness and the Governance Agenda." In *Government at a Glance 2013*. Paris: OECD Publishing.

Tan, Khee Giap, Randong Yuan and Wei Cher Sangiita Yoong. 2015. *2014 Annual Provincial and Regional Competitiveness Analysis for Greater China Economies: Development Strategies under the New Normal*. Singapore: World Scientific Publishing Company.

Tan, Khee Giap, Randong Yuan, Wei Cher Sangiita Yoong and Mu Yang. 2014. *Annual Analysis of Competitiveness, Simulation Studies and Development Perspective for 34 Greater China Economies: 2000–2010*. Singapore: World Scientific Publishing Company.

Veblen, Thorstein. 1915. *Imperial Germany and the Industrial Revolution*. New York: Viking Press.

Wu, Min. 2007. "邓小平的政治体制改革蓝图及其实践 Deng Xiaoping's Blueprint and Practice for Political Reform (in Chinese)." *炎黄春秋 Yanhuang Chunqiu* 7.

Ye, Jingzhong, James Murray and Yihuan Wang. 2005. *Left-behind Children in Rural China: Impact Study of Rural Labor Migration on Left-behind Children in Mid-West China*. Beijing: Social Sciences Academic Press.

Zou, Dongtao, ed. 2008. *Blue Book of Development and Reform (no. 1): China's 30 Years of Reform and Opening-up (1978–2008)*. Beijing: Social Science Academic Press.

Discussant Notes 1

A Review of the "2015 Annual Update on Competitiveness Ranking and Simulation Studies for 34 Greater China Economies" and "2015 Regional Competitiveness Analysis for Greater China"

Dr He Dongni

Vice President, China Institute for Reform and Development

People's Republic of China

1. GENERAL OPINION

The comprehensive report on the 2015 Provincial and Regional Competitiveness Analysis for Greater China Economies has three salient features.

First, it has covered beyond a single target of GDP growth. The comprehensive list of competitiveness indicators includes four major environments, 11 sub-environments and 102 indicators, which are closely linked with the guiding principle of achieving sustainable development.

Second, it is a piece of policy-oriented research. Through the horizontal comparison of competitiveness among the 34 economies and five regions and the comparison of the competitiveness score for each economy or region against the median and maximum scores, the report offers a clear diagnosis on the potentials, shortfalls and problems of each economy or region, thereby proposing targeted policy recommendations.

Third, it contains a lot of valuable information. For example, the report analysed the changing trends in economic influence by the United States (US), the European Union (EU), Japan and China to the five countries in ASEAN, which is a very valuable and enlightening piece of reference for the study of regional integration and regional governance structure.

2. Issues that Need Further Discussion

Regarding the entire research framework, the framework of the comprehensive competitiveness study is undoubtedly systematic and innovative. To further improve the adaptability and operability of the research framework of this study, there might be a few questions that need to be answered:

First, in the research framework, is there a middle level of competitiveness analysis that is between the micro and macro levels, i.e. covering the level of industries or sectors of the economy? From the four environments and 11 sub-environments of indicators, the environments of Macroeconomic Stability and Government and Institutional Setting cover more on the macro level, perhaps some indicators under the sub-category of Institutions, Governance and Leadership are micro level indicators on firm behaviours in the markets. But looking at the surface, these two environments cover more on the macro level. Indicators in the environment of Financial Deepening and Business Efficiency seem to be based on considerations for the middle and micro (firm) levels, but it is not very clear. So, in which environment do we cover the middle-level competitiveness analysis? Among the 102 indicators, are there any indicators assessing the competitiveness of the real economy and pillar industries of a country or region or economy? If so, what is the situation like for the middle-level competitiveness among the 34 Greater China economies?

Second, the last environment among the four environments is Quality of Life and Infrastructure Development, but when analysing competitiveness in this study, why do we put the quality of life and the infrastructure development together? What consideration is such a way of organising indicators based on?

Third, in this research framework, with "Quality of Life" being used as an important measure of the Overall Competitiveness, is the "Quality of Life" here the result or the source of competitiveness? This issue is very important, as China is in a crucial period of transitioning to a consumption-led economy. We know that the consumption level and quality of consumption are important indicators for the standard of living and quality of life. In ACI's research framework for the Overall Competitiveness, is there any consideration on the effects of consumption (consumer market) on the Overall Competitiveness?

Recently, I read an article saying that China is the only country that has established a complete range of secondary industries in the world, and that we are 100% complete while the US is not 100% but above 90% complete. One of the most important reasons for why China can do it is that with 1.3 billion people China has a consumer market that is big enough and various levels of consumer demands that are diverse enough to support such a large system of secondary industries.

Since 2008, when the Global Financial Crisis broke out, President Chi Fulin of China Institute for Reform and Development has repeatedly emphasised the need for the Chinese economy to transform into a consumption-led growth pattern. In the last three years of research, President Chi and our research team have become more certain about one point that the key for whether China can successfully escape the middle-income trap and overcome the small challenge in 2017 and the big challenge in 2020 lies in whether China can seize the opportunities in the restructuring and upgrading of secondary industries, restructuring and upgrading of urbanisation, and most importantly, the upgrading of the consumption structure to form a services-dominated industrial structure.

In addition, there are several conclusions in the analysis that somewhat confused me, which may be further discussed.

First, Liaoning is ranked in the top 10 among the 34 economies in terms of the Overall Competitiveness (its ranking improving by one position in 2015 over 2014, overtaking Macau to be ranked at ninth). Why the confusion? From GDP, growth of secondary industries and other indicators, the current situation in Liaoning should be considered as a rather severe case of the "Northeast Emergency", and Liaoning's secondary industries value added decreased by an annual rate of 4.5% in the period from January to February this year according to a statistic I saw, which is the worst province in China, with only two provinces in the country experiencing negative growth and the other negative growth province being Shanxi. I am not quite familiar with the situation in Macau. Mao Zedong put forward in his article "Against the Doctrinism" in May 1930 that "no investigation, no right to speak". I did not do in-depth research, so I am not speaking but asking. Professor Chang Xiuze has just come back from the Northeast with a team of PhDs doing in-depth research all over the region there. With revitalising the Northeast being such an enormous challenge, I would like to hear from Professor Chang for his views on the ranking.

Second, Fujian is ranked as the top of the 14 middle-ranked economies (obtaining a ranking of No. 11 for both 2014 and 2015, after Liaoning and Macau). Fujian belongs to the Eastern Coastal Area. Of course, that does not mean that the economies from the Eastern Coastal Area are necessarily competitive. But no matter it is regarding the industrial restructuring and upgrading or the potential for economic growth driven by the openness of the economy, it should be said that Fujian is relatively strong. So I am also confused by the ranking of Fujian.

Third, Chongqing and Jiangxi are ranked in the bottom 10 among the 34 economies in terms of the Overall Competitiveness (Chongqing belonging to the Western China and Jiangxi belonging to the Central China). I am not that confused by the horizontal positions of the ranking, but quite confused by the longitudinal changes in the ranking. Chongqing is ranked No. 26 in 2015 for the Overall Competitiveness, falling by one position as compared to 2014. If judged only by the perspective of economic growth, the growth rates of Chongqing for the first half of the year in 2013, 2014 and 2015 are all among the top three fastest of the country, indicating the clear trend of the growth potential being released rapidly. Then it was overtaken by Yunnan in 2015. What are the key factors for such change in the ranking? Jiangxi is ranked No. 27 for the Overall Competitiveness for both 2014 and 2015. But its ranking in the Macroeconomic Stability has declined by two positions (from No. 21 to No. 23), and its ranking in the Quality of Life and Infrastructure Development has fallen by three positions (from No. 27 to No. 30). The changes in the rankings for the environments do not seem to be reflected in the ranking changes in the Overall Competitiveness. I hope these questions and confusions may stimulate our further discussion.

3. Inspirations and Suggestions for Research

Generally speaking, ACI's comprehensive research framework for studying competitiveness gives us a good perspective to examine and analyse the development of an economy, from which

I am very inspired. But to look at it from a deeper level, I have three immature research suggestions.

First, in order to let the local governments accept and consciously make use of the indicators for the Overall Competitiveness to assess the local development, we must give a clearer answer as to why the Overall Competitiveness is so important now. After 30 years of the reform and opening up, what kind of great practical significance is there if the local governments now compete in improving their Overall Competitiveness rather than compete in increasing their GDP?

Second, we should make sure that the entire research framework is based on a system of indicators that can stand the tests of time and practice, which means that the system of indicators needs to constantly update and develop itself. As we all know, no matter it is for a country or a region, a proper treatment in the relationship between government and market and the relationship between government and society is crucial to their development. For example, how much is the weight for the indicators reflecting the strength in the institutional systems? Can they well reflect the effect of the relationship among government, market and society and the corresponding institutional arrangements?

Third, we should improve on policy recommendations with deeper and more solid research. In order to reach the goal of providing advisory services for policy decisions for the 34 economies, it is necessary to make targeted policy recommendations based on more detailed analysis of the many indicators. For example, as in the previously mentioned cases for Liaoning, Fujian, Chongqing and Jiangxi, why has the ranking not changed for some of the economies with improved rankings in terms of economic growth rates? Which indicators are dragging them down? Which indicators are their weaknesses? How to make up for the weaknesses? How to promote positive changes in slow variables among the indicators? These four economies are located in North-eastern China, the Eastern Coastal Area of China, Western China and Central China. They are each quite representative of the situation faced by their respective regions in the entire economic transformation of China, which are worthwhile candidate sample set of economies to be studied with more in-depth and specific research efforts.

4. Three Thoughts on the Main Themes

In order to improve the Overall Competitiveness of provincial level regions of China (including the municipalities, autonomous regions and special administrative regions), we may need to do the following.

First, in the process of restructuring economic development, each place across the country should adapt to and capitalise on their local conditions without rushing together in the same direction, and they should foster the development of their own competitive advantages based on their own characteristics. We have learnt deep lessons from the excess capacity caused by the lead in economic activities by local governments. Therefore, with regard to the economic transformation and upgrading during the "Thirteenth Five-Year Plan", the most important thing is to seek a proper treatment in the relationship between government and market and the relationship between government and enterprises without using industrial policies to distort market behaviours.

Second, structural reforms are the common theme. No matter it is regarding the Northeast Area Revitalisation Plan, the Rise of Central China Plan, or the policies in "Emptying the Cage for the New Birds" in the Southeast coastal area, in order to realise them, we need to carry out a package of structural reforms in areas such as market institutions, fiscal institutions, financial system and education system.

Third, the local governments should put themselves in the right position in the overall plan of China's opening up. For some regions, such as Hainan, opening up is absolutely necessary, because it is an island economy and indeed located at a relatively important position in the Maritime Silk Road for international cooperation. But if all the provinces or areas rush together into speculating on the concept of "One Belt, One Road" and launch new industrial projects and construction plans using "One Belt, One Road" as a pretext, it might stimulate investment demands in some places, but such "boost" to the economy is not sustainable and will likely lead to an unpleasant eventuality.

Discussant Notes 2

A Review of the "2015 Annual Update on Competitiveness Ranking and Simulation Studies for 34 Greater China Economies" and "2015 Regional Competitiveness Analysis for Greater China"

Professor Zhang Shucun

Dean, Shandong Academy of Social Sciences
People's Republic of China

Just now, from different perspectives, Mr Yuan Randong and Ms Sangiita Yoong presented the latest results for the 2015 Provincial and Regional Competitiveness Analysis for Greater China. These results are distilled from empirical data based on a research framework consisting of four environments, 11 sub-environments and 102 indicators. Through rigorous technical means, the provinces and regions of Greater China were comprehensively assessed and the rankings according to their Overall Competitiveness as well as the four environments that make up the Overall Competitiveness were released. The rankings of the five regions of Greater China were also presented.

Competitiveness is a very complex issue, which involves many aspects such as the economy, the society and the culture. The excitement of the media and the public usually lies in the ranking sequence of various kinds of competitiveness rankings. Scholars may be interested in the choice of indicators and whether the research methods and data sources are scientifically treated. Government officials including our think tanks are more interested in the policy implications behind the rankings. I think that the system of the indicators used in this report is constructed in a reasonable manner, the adopted methodology is also scientifically designed, the analysis is rich in content, the data used are credible, the division of the large regions is based on China's national and regional strategies, and the results of the analysis basically reflect the current actual situation for the development of the provinces and regions of Greater China, which embodies the characteristics of the study of being objective, impartial, independent and being conducted on a *pro bono* basis, with some of the proposed judgments also being innovative. I feel this is a piece of high quality work from a think tank.

Furthermore, the report contains the analysis of the latest regional competitiveness rankings, the characteristics of the ranking changes and the intrinsic connections and associations, which forms an important reference and provides policy recommendations for enhancing the competitiveness of the regions. The sections related to mainland China have a positive reference value for drafting the "Thirteenth Five-Year Plan" and promoting a comprehensive, coordinated and sustainable development for the economy and the society of the regions. I have learned a lot from this report. I would like to take this opportunity to extend our warm congratulations to the experts from the Asia Competitiveness Institute, to the Co-Directors Professor Tan Khee Giap, Professor Tan Kong Yam and the research team on behalf of the Shandong Academy of Social Sciences.

The review seminar today is not only the platform for the Asia Competitiveness Institute to promote its think tank products, but also a platform for the exchange of academic achievement. The Shandong Academy of Social Sciences is implementing innovative social science research projects and engaged in building the new types of think tanks with academic support. I also have personal experience in public services related to government decision-making. Using this opportunity, I would like to raise a few questions and seek academic exchanges with the experts.

First, it is regarding the judgment on the development trends in the regional spatial patterns in mainland China. In today's report, mainland China is divided into the four regions, namely the Eastern Coastal Area of China, Central China, Western China and North-eastern China. This is a widely recognised traditional way of defining the regions, which has its rationality. However, China's regional strategy is currently being adjusted, shifting from blocks of regions as the main consideration to a combination of blocks of regions and the axes or lines being the main consideration. In the process of improving the evaluation system, we should gradually take into account this factor. Meanwhile, in view of the development trends, the borders between the regions or blocks will become increasingly blurred. Aggregation, integration and inclusive development are the basic characteristics of future regional development.

The report also refers to the impact of transport infrastructure on competitiveness. Various studies show that in 2010 or earlier the investment in transport infrastructure could exacerbate inter-regional imbalances in economic growth and development. This is called the Matthew effect. However, according to the research findings by Professor Ni Pengfei who is present here today, with the development of high-speed railways, the contact and exchanges between the regions are rapidly becoming increasingly frequent. A close cooperation among the Eastern Coastal Area, Central China and Western China has been taking place. Together with the impact of the new national urbanisation strategy and the "One Belt, One Road" strategy, the level of aggregation and integration of the four regions will steadily increase. In this context, there will be profound changes in the patterns of the demarcation of regions which are now defined as the Eastern Coastal Area, Central China, Western China and North-eastern China. We should strive to discover and reveal new patterns and future trends as reflected spatially or in terms of the evolution over time, and demonstrate them clearly in the system of indicators.

Second, in the new model of development driven by innovation and the new generation of information technology, we have just had a lively discussion on the issue of innovation and the distinction between manufacturing and smart manufacturing. As an important part of the competitiveness analysis, scientific and technological innovation and informatisation should be studied with more rigour.

Third, for assessing the Overall Competitiveness, there are some indicators reflecting quality of life, standard of living, education and environment. We did not see the specific indicators today but it should be said that the selection of these indicators is extremely complex and difficult, especially for indicators related to subjective perception and personal experience. We should be extremely cautious about the judgment on the weights and calculation methods. In addition, under the current situation, some of the indicators on public services, such as high quality educational resources, medical resources, business environment and the impact of the culture on economic development and so on, have a great impact on the potential to enhance competitiveness, which will become increasingly important. This means that the government should provide better public services and better basic conditions for human existence.

Finally, the report has proposed three policy recommendations after providing the provincial and regional rankings. I think that these policy proposals really capture the key issues. But the circumstances vary widely across the regions and provinces, and we should constantly explore the policies on the operational level in order to make them more targeted. In short, this report is of very high quality and importance. As peer think tank researchers, we will study the details of the report carefully.

Discussant Notes 3

A Review of the "2015 Annual Update on Competitiveness Ranking and Simulation Studies for 34 Greater China Economies" and "2015 Regional Competitiveness Analysis for Greater China"

Dr Ni Pengfei

Director, Centre for City and Competitiveness
Chinese Academy of Social Sciences
People's Republic of China

I would like to first congratulate on the release of the report. I noticed that this is a very young research team, and they did a lot of in-depth research. I want to talk about three points. The first is a positive appraisal for the report. The report is a good piece of work for three reasons.

First, I think this is a very good topic. It should be said that China is a big country. Research on China at the national level tends to mask some regional issues. So the study of provincial competitiveness is very necessary.

Second, it has very important impacts. In my previous studies on competitiveness, I found that it has three impacts. The first impact is generating a competitive incentive. Once ranked, we will compete to be ranked higher. The second impact is playing a guiding role. If your system of indicators emphasises on this indicator, your research subjects will pay attention to this aspect. The third impact is serving as a reference for policy decision-making. This is its basic function. As for the second impact of the competitiveness report, with the Asia Competitiveness Institute from the National University of Singapore studying the issue of China's regional competitiveness, it has above all an independent identity, and secondly it has an international perspective. This is a very critical assessment factor. Just now we have discussed how to achieve independence and how to achieve objectivity. I believe that the funding they receive is not from any provincial government from mainland China, which is very important.

Third, there are some points in the conclusion that I think are very good. For example, the comparison with the median and maximum scores as well as the What-If Simulation Analysis are all very useful research methods. At least we have not made good use of them in the past. As far as I am currently aware of the research results related to the overall strength of the economy, the development level as well as the competitiveness level, it should be said that the provincial competitiveness rankings in this report are largely in line with them. This is the positive side.

There are, however, several questions to be discussed. The first is the framework. The theoretical basis of the competitiveness research framework should be further clarified. There are generally three theories on competitiveness. The theory on the long-term economic growth is a basis. Another basis is the inter-regional and international trade theory. Of course, now we are using a macroeconomic framework to carry out our research. What is your theory? How does each indicator match your theoretic framework? At least I am not very clear about that after reading the PowerPoint slides.

The second question is the one I particularly want to emphasise on. You now have four environments, with each environment being a sum of two aspects. For example, Financial Deepening and Business Efficiency, Quality of Life and Infrastructure Development, Government and Institutional Setting. After you synthesise them, you would not be able to say what they are. This is a very critical issue, which I hope to draw your attention to. You put two unrelated things together and label them with the word "and", but what is that indeed? I think that this is a very important question.

Thirdly, no matter what kind of competitiveness, I think the core issues are talents, technology and education, which should be a focus of the study. Of course there are indicators related to them in the framework, but how to make them more prominent?

The fourth question is that Central China is ranked at the bottom in the regional competitiveness results of the study, which I think is debatable. Let's have a contrast between our research results and yours later. In terms of policy recommendations, since the team members are all very young, just beginning to do research on China, so it is understandable that recommendations are still not targeted enough.

We have done some research for the case of Central China. Our report on Competitiveness of China released this year is focused on the study of the problem of the spatial structure, which also makes observation from the provincial and regional perspectives. China is now into the "Thirteenth Five-Year Plan", with particular emphasis on the optimisation of the spatial structure of the economy. We found that the spatial structure of China, if carefully studied, is not as divided into the Eastern Coastal Area, Central China, Western China and North-eastern China as people may have imagined. According to your provincial data shown just now, it is clear that the Eastern Coastal Area is all in good shape and the situation of Central China is not that great. But if you look at NASA's night lights map, you can see that it is not the case that the whole block of the Eastern Coastal Area is strong economically. It is actually the major cities and clusters in the Eastern Coastal Area that are very strong. Some places in the Eastern Coastal Area are as underdeveloped as many parts of Central China. And there are quite advanced areas in Central China as well.

We noted that the Eastern Coastal Area and Central China have formed a network-like structure. We also noted that the Heihe–Tengchong Line continues to exist. But we have found another important line, which is the 110°E Line. This is my discovery. I found that the 110°E

Line serves as a great fault dividing the east and west in an economic sense. Areas that lie on the east of this line are quite similar to the Eastern Coastal Area in terms of economic vibrancy. We can further examine it using the economic density data, which offers a very similar conclusion as per the 110°E Line. Of course, the 110°E Line is more accurate, which becomes even clearer from the changes in economic density from 1990 to 2010.

The above discussed are the observation of some phenomena. We now face another very important factor, which is that China has entered the era of high-speed railway economy. High-speed railway economy has changed the scale for spatial distances, making what was measured originally in spatial scales to become measured now in time, thereby making it possible to produce the effect of compressing space. By 2030, the high-speed railways in China as a whole will form a network of at least six vertical lines and six horizontal lines in the Eastern Coastal Area and Central China. The time dimension will be compressed with the speed increasing from 100 kilometres per hour to 300 kilometres per hour.

Based on such considerations, we have made an important finding. In terms of the regions, there may be a significant change to the four regions in the future. The Eastern Coastal Area and Central China may merge together to form the core of China, with the Northeast and Southwest of China becoming the peripherals, forming a structure of integrated Eastern Coastal Area and Central China with tilted peripheries. Such a conclusion on the future structure of China is mainly based on population density. Supporting this spatial structure is the urban systems of China, which is because we are in an era of urbanisation. For the urban systems, generally speaking, they will be structured as one group and five lines. It means that there will emerge a spatial structure of networks that consists of high-speed railway networks, highway networks and so on in the Eastern Coastal Area and Central China. There will emerge five lines in the Northeast, Northwest and Southwest of China according to the conditions in transportation development, which will happen in coordination with the opening up of China and the "One Belt, One Road" strategy. This will be the situation of the future, to which we of course made a number of important policy recommendations. This is our research, which also can provide some reference for you.

Chapter 4

Agricultural Productivity in Greater China Economies: Estimating Provincial Agricultural Total Factor Productivity in Mainland China

4.1 MOTIVATION AND EXPECTED CONTRIBUTIONS

In the era where avant-garde activities of the global economy are characterised by human endeavours in areas such as biotechnology, space exploration, cloud computing, big data and artificial intelligence, agriculture as an issue so down to earth may not seem to be that exciting a topic. This is especially so for China, with a dual identity of being a fledgling major power but at the same time a developing country still. Why should people both inside and outside of China care about the development trends in Chinese agriculture and its productivity? The short answer is that it matters to them in fundamental ways.

To put the issue of Chinese agriculture in a more alarmingly tangible context, China has to feed one-fifth of humanity,[1] or its people will face the threat of starvation. This, in turn, might derail China's economic development *en masse*, and people in other parts of the world and neighbouring countries in particular might have to deal with the risk of large inflows of refugees as the unfortunate recent events in the Middle East and Europe have demonstrated. The resources at the disposal of the country to accomplish this task are rather limited; China possesses less than 10% of the entire agricultural land of the world and less than 6% of the total availability of fresh water (Tan, Suo and Rajan Forthcoming).

The migration of people from rural to urban areas and the expansion of cities and towns along with rapid conversion of agricultural land into industrial zones all across China, possibly the largest and fastest urbanisation process to ever happen in human history, also serve to further strain the supply of agricultural land and shift manpower away from the primary sector. Thus,

[1] At the time of writing this book, the population of mainland China ticked at 1,374,140,337, according to China's population clock provided by the National Bureau of Statistics of China website, accessible at http://data.stats.gov.cn/, while the world population reached 7,296,965,107 according to the data provided by the United States Census Bureau, accessible at http://www.census.gov/popclock/. The size of population in mainland China is thus approximately one-fifth of the world population, or 18.8% to be more precise.

the two major inputs of the production function for agriculture, namely labour and land, offer no reassuring outlook as long as the certainty of growth in agricultural products is of concern; the number of farmers is actually dwindling, while the acreage of farmlands stagnates to say the least.

All these necessitate a significant improvement in agricultural productivity. Against such a backdrop, achieving and sustaining sufficient level of agricultural productivity is paramount if China is to ensure food security in the coming years. Ensuring significant productivity growth in agriculture will also help raise farm wages, hence enhancing the welfare and standard of living of rural dwellers and providing them with the incentives to continue to work and invest in the agriculture sector. Therefore, in this chapter, we choose to study agricultural productivity of China and will pay special attention to analysing the Total Factor Productivity (TFP) for China's provincial level economies.

The objective of our study is to examine agricultural productivity growth at the provincial level in China with a time series dataset covering the last two decades. We will take a non-parametric approach to measure agricultural TFP growth in China among the provinces. Such a methodology helps to break down the change in TFP into two components, namely the technological change (TC) and the technical efficiency change (TEC), which relate broadly to the progress in technology (i.e. shift in the production frontier) and better resource utilisation through more efficient management, respectively.

This research is driven by academic interests that serve as a modest spur to induce healthy debate and further research on this seemingly low profile but critical topic. The results of such empirical investigation will enable us to have an updated understanding of the key factors that determine the growth profile of the agricultural sector at the subnational level in China, thereby helping relevant stakeholders from the public and private sectors to make more informed decisions.

Before delving into the details of the empirical analysis and estimation, we will briefly discuss the development of agriculture since the founding of the People's Republic of China, before and after the reform and opening up, with special emphasis on the past two decades.

The development of agriculture has seen some mixed performances in the three decades (1949–1978) before the implementation of economic reforms in China. Overall production had increased, with the production of staple food such as grains rising by more than 2.5% annually from 1952 to 1978, outpacing the average annual growth rate of population over the same period of 1.9%, indicating the generally improved situation for staple food availability (Huang and Rozelle 2015). However, the supply of the variety of food items beyond staple food remained rather limited in the country except for very big cities of strategic importance in the period before the reform. During this period, agricultural productivity gauged in measures such as TFP did not make significant gains (Wen 1993), and as a result farmer income seldom moved upwards (Lardy 1983).

The increase in agricultural output was largely attributable to government-directed investment funded through taxes and partially supplemented by underpaid and sometimes unpaid labour from the farmers (Huang and Rozelle 2015). Meanwhile, the rigid pricing system under the command economy era did not help induce the right incentives for more efficient production and distribution of agricultural products. The growth in agriculture before the reform and opening up basically lacked the vitality and dynamism required for any economy poised to take-off.

The development of agriculture in China swiftly changed to a higher gear since the onset of the economic reforms in 1978, with entirely another level of growth momentum. The Gross Value of Agricultural Output of the country had registered an annual growth rate of 7.1% in the first six years since the reform (1978–1984), an increase in growth rate of almost 200% as compared to the prior years. From 1985 until 2010, the annual growth rate of Gross Value of Agricultural Output maintained on average at about 4%, a very respectable level by any international standard for such an extended period (Huang and Rozelle 2015). The growth has spread widely in almost all sub-sectors of agriculture in China, with major diversification and increase in agricultural output beyond staple food such as grains, as exemplified by the growth in the production of livestock and fisheries being about twice as fast as the growth in crops in the past three decades since the reform (National Bureau of Statistics of China 1981–2012).

Reassuringly, the dynamics of agricultural productivity after 1978 did not follow the flat pattern previously seen before the reform. Since the beginning of the 1980s, the yields in agricultural output per unit of land or per unit of labour experienced significant rise (Huang and Rozelle 2015). During the same period, there were also pronounced positive changes in TFP, according to many researchers whose works will be discussed later in Section 4.2. Accompanying the rise in agriculture productivity is the sharp increase in farmer income, achieving an average annual growth rate of 7.5% from 1978 to 2011 (Huang and Rozelle 2015), which is both important and encouraging as it attests to the principle that the well-being and happiness of the people should not be deemed as subservient to any economic system or political regime.

It is worth noting that one idiosyncrasy stands out in the organisation of agricultural production in China. After more than 30 years of reform, the lion's share of economic activities performed in Chinese agriculture are still carried out by individual families controlling and managing small farms; the average agricultural land area per rural family hardly increased from 0.70 hectare in 1980 to 0.75 hectare in 2010 (Huang and Rozelle 2015). Regarding how actively the cooperative farming organisations operate, such cooperatives were found in only 20% of villages in China in 2008, with only 10% of all farmers in China being affiliated under any of them (Deng *et al.* 2010). This is in sharp contrast with quite a few East Asian economies and almost all of the developed countries, where the majority of farmers participate in some form of agricultural cooperative organisation (Huang and Rozelle 2015).

Some observers pointed out that such an arrangement based on small family-run farms actually helped to maximise gainful employment of rural residents and therefore contributed to the increased total agricultural outputs (Studwell 2014). While such arguments might be true in the initial phase of China's economic transformation, when urbanisation and industrialisation did not unfold in earnest, such fragmented organisation will nonetheless be bound to encounter bottlenecks due to its lack of scale economies. However, such a situation of the Chinese primary industry dominated by individual and small-scale farmers may just be a blessing in disguise, as it could potentially offer ample scope for further reforms to bring about sizeable productivity improvement, especially in one component of the TFP that is the TEC as mentioned previously (this will be discussed in greater details in the later sections of this chapter).

Figure 4.1: Primary Sector* Value Added and Employment

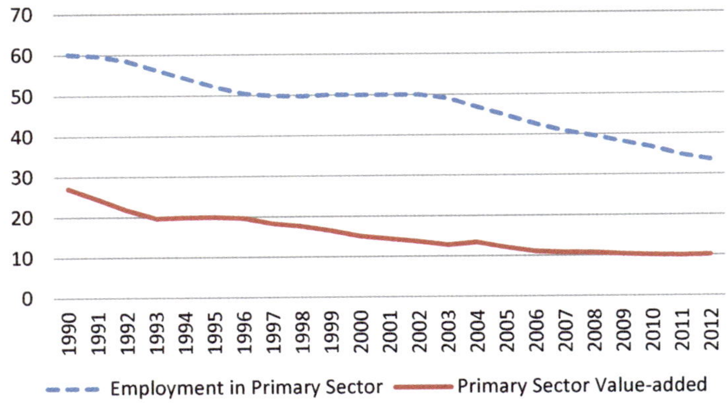

**Note*: According to *China Statistical Yearbook*, primary sector includes farming, fishery, forestry and animal husbandry
Source: China Statistical Yearbook 2013

It is quite plausible that the rise in agricultural productivity in China in the early period of the reform is attributable mainly to factors related to institutional and agriculture policy reforms involving incentives, pricing and marketing mechanisms (Lin 1992; Fan and Pardey 1997). But just like launching a multistage rocket into space, the quest for betterment in the development of agriculture in China should have gone beyond the first stage boosted by institutional and policy reforms to dismantle the communes (De Brauw, Huang and Rozelle 2004). From here on, moving upwards will increasingly require technological breakthroughs in agriculture generated by research and development (R&D) activities as well as better dissemination and application of existing technologies and best practices.

Hence, we choose to focus on the data from the recent two decades to incorporate the most recent development trends in agriculture in China as well as to avoid possible structural breaks bound to exist if the timeframe chosen was overly lengthy. We hope that our study can serve as an attempt to fathom how these up-and-coming factors for boosting agricultural productivity would have played out in recent years, therefore offering a base upon which to postulate the likely courses about the near future and the corresponding best options for responses.

The primary sector (i.e. agriculture sector, including farming, fishery, forestry and animal husbandry) as a share of Gross Domestic Product (GDP) and total employment has been declining over the last two decades. As seen in Figure 4.1, the primary sector's contribution to national output had fallen from 27% in 1990 to 10% in 2012. During the same period, its contribution to national employment decreased from 60% to 37%. The decline of agricultural output, however, only happened in relative terms as seen in its percentage in the national output. In absolute terms, the primary sector had increased steadily over the same period at an average annual real growth rate of 4.4%,[2] which underpins the fact that the standard of living for people in China improved dramatically. The drop of agriculture in relative terms and its increase in absolute terms merely

[2] Calculated by authors based on the data from the website of National Bureau of Statistics of China at http://data.stats.gov.cn/.

Figure 4.2: Composition of the Primary Sector in 2012 (%)

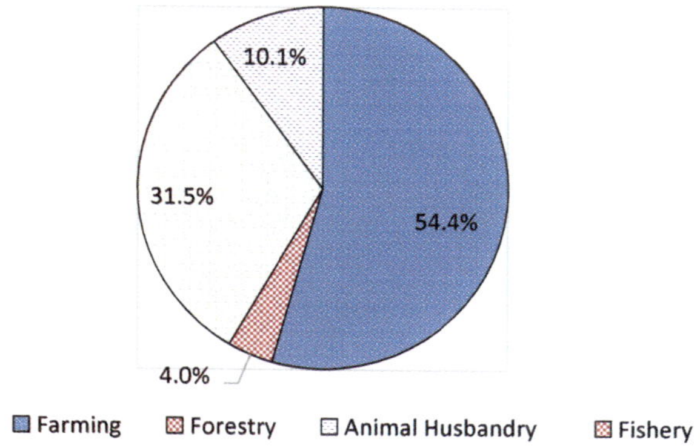

Source: China Statistical Yearbook 2013

constitute a natural reflection of Engel's law.[3] In contrast, the number of people employed in the primary sector actually declined gradually from 1990 to 2012, at an average annual rate of −0.93%.[4]

China has various kinds of climate, land relief and soil conditions, some of which are conducive to developing various agricultural activities, such as animal husbandry, aquaculture, cash crops farming, forestry and grain crops farming. Figure 4.2 shows the composition of the primary sector in 2012, in which farming constitutes more than half of the primary sector's output values, followed by animal husbandry contributing almost one-third of the total output. Fisheries and forestry in total contributed 14% to the total output. As mentioned previously, the growth of fisheries and animal husbandry outpaced the growth of staple food farming in recent decades, with their shares in the total agricultural output values likely to increase in the coming years.

Despite its declining share in China's overall economic activity, agriculture is still a fundamental source of income for a substantial portion of China's population, providing the fresh produce used to make the majority of food items consumed by its people[5] and the raw products as indispensable inputs to fuel the growth of its industries.

Examining agricultural trade patterns in China, one can observe that after experiencing some volatility in the 1990s, agricultural imports became stable since 2000. Meanwhile, the corresponding share of agricultural product exports in total merchandise trade experienced a significant decline between 1990 and 2012 (Figure 4.3). China has been a net agricultural product importer since 2000, as shown in Figure 4.4. Apart from the growing share of agricultural imports to total merchandise imports, the value of agricultural imports increased faster than the value of exports during the same period post Global Financial Crisis (GFC), implying an increase in the net agricultural import value. This underlines the need for China to enhance its

---

[3]Engel's law says that the poorer the household, the higher the share of its expenditures spent on food.

[4]Calculated by authors based on data from China Statistical Yearbooks, China Rural Statistical Yearbooks and Provincial Yearbooks.

[5]As of 2010, China was still mostly self-sufficient in raw and processed food, reaching a self-sufficiency level of 97% (Huang and Rozelle 2015).

Figure 4.3: Share of Agricultural Trade to Total Merchandise Trade (%)

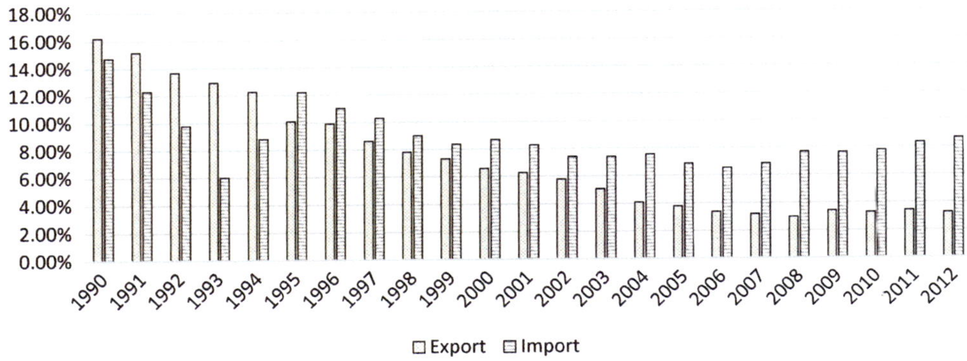

Source: World Trade Organization (WTO)

Figure 4.4: Value of Agricultural Trade 2009–2012 (Billion US$)

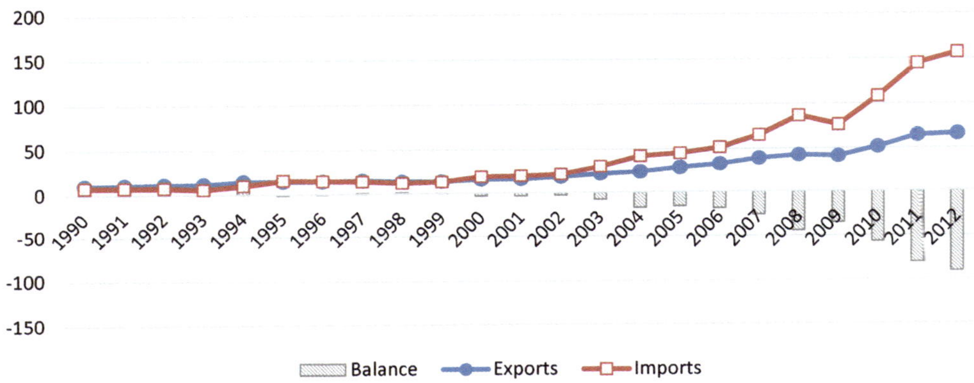

Source: World Trade Organization (WTO)

agricultural productivity to reduce its dependency on overseas agricultural products and thereby strengthen its domestic agriculture sector.

Given China's large size and heterogeneity, it is important to move beyond the national level and examine data at the provincial level. Figures 4.5 and 4.6 illustrate the provincial primary sector output value and its share of the Gross Regional Domestic Product (GRDP). The top six provincial level producers, *viz*. Shandong, Henan, Jiangsu, Sichuan, Hebei and Hunan account for almost two-fifths of total primary sector output in China. Ningxia, Qinghai, Tianjin, Beijing, Shanghai and Tibet are the bottom six provinces, accounting for less than 2% of the primary sector output of China. In terms of the share of primary sector of GRDP, there has been a decline across all the provinces between 1992 and 2012. Tibet, Inner Mongolia and Jiangxi experienced the steepest decrease, while Liaoning and Heilongjiang experienced the lowest decline during the period, which is consistent with the phenomenon of the relatively sluggish pace of further

Figure 4.5: Provincial Primary Sector Output in 2012 (100 Million Yuan)

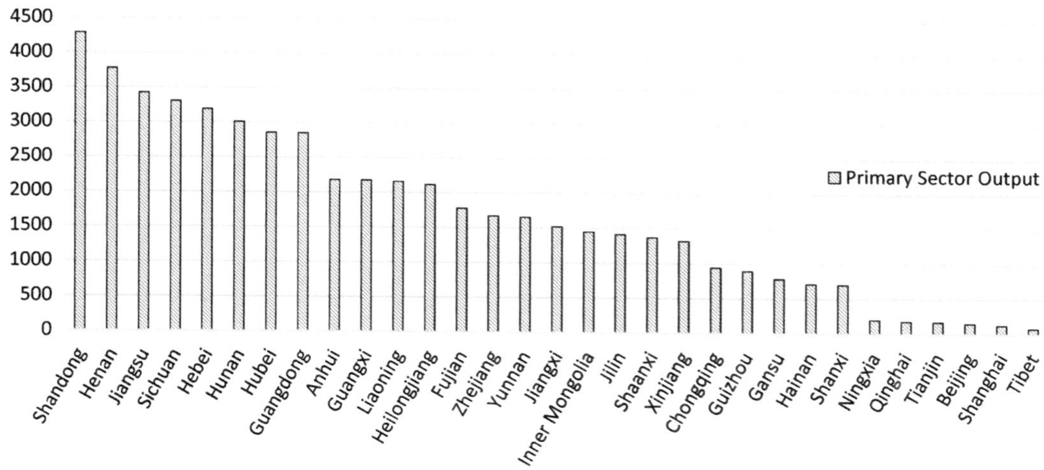

Source: China Statistical Yearbook 2013

Figure 4.6: Provincial Share of Primary Sector to GDP (%, 1992[6] and 2012)

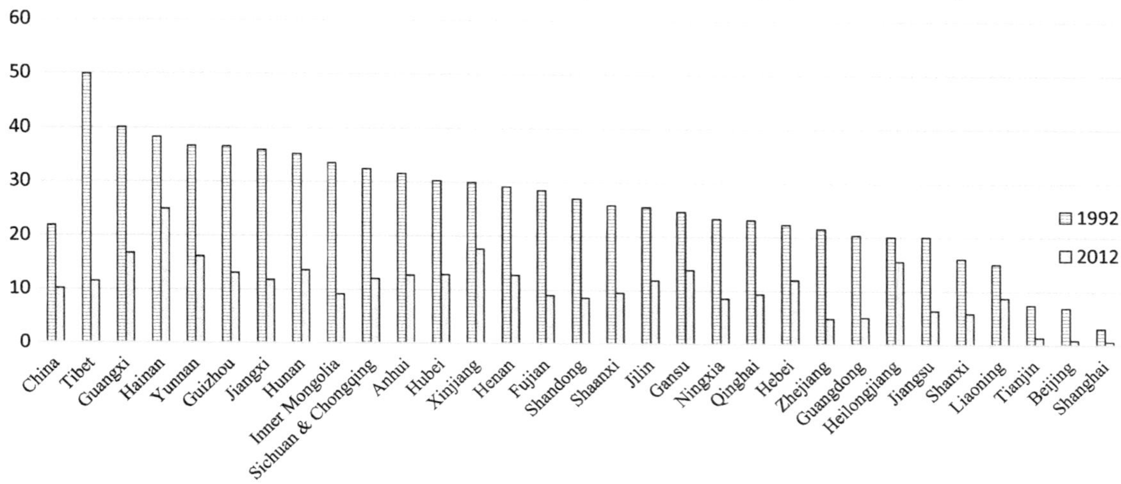

Source: China Statistical Yearbook 1993 and 2013

industrialisation and economic growth as observed in the North-eastern provinces compared to other regions of mainland China. This also has to do with the fact that Liaoning and Heilongjiang, as the provinces where many of the North-eastern old industrial bases are situated, used to have one of the highest industrialisation levels in China. Therefore, they had a lower starting point for the share of agriculture in GRDP to begin with (Tan, Yuan and Yoong 2015).

[6]China adopted GDP statistic at 1992. Before 1992, the output value was measured by the overall society output value.

It is worth noting that, considering the wide variation in agricultural production among provinces, our interest in this chapter is to analyse agricultural productivity at the provincial level. In addition, we also focus on trying to understand agricultural TFP growth and its driving factors, as well as provide suggestions to improve provincial TFP.

Based on the literature, there is a perception that technological progress has been the key driver in China's agricultural sector in recent decades. To verify this hypothesis we apply growth accounting and the Malmquist index Data Envelopment Analysis (DEA) methodologies to estimate the annual change of TFP for mainland China provinces for the period 1990–2012. To preview the main conclusions, while our empirical results confirm the importance of technological progress in TFP growth for most provinces, it must be emphasised that efficiency improvements turned out to play a bigger role for some provinces. Our results also provide policy insights for local governments to develop targeted strategies to enhance agricultural productivity.

The rest of the chapter is organised as follows. Section 4.2 discusses the previous studies on agriculture of China conducted at the national and provincial levels. Section 4.3 presents the empirical approach for measuring TFP and the corresponding data description notes. Section 4.4 discusses the results of the study. Section 4.5 concludes by highlighting selected policy implications.

4.2 LITERATURE REVIEW

Some of the literature on agricultural productivity for China to date is summarised in Table 4.1. We briefly review some of the main literature below.

There have been a number of studies on TFP in China's agricultural sector at both the national level and provincial level. Fan and Zhang (2002) amended the previous measure of livestock and fishery output and calculated a Törnqvist index to measure the TFP at the national and provincial level during 1952–1997. They found that the TFP annual growth rate was 3.28% between 1979 and 1997.

Mead (2003) revised China's labour force data and calculated the TFP growth for Chinese agricultural productivity based on a constant-returns-to-scale Cobb–Douglas methodology for the period of 1982–1999. The results showed a strong correlation between the changes of agricultural policies and agricultural factor productivity growth in this period.

Tong, Fulginiti and Sesmero (2009) argued that data from *China Statistical Yearbook* showed the fluctuation of output value and inputs, while data from the Food and Agriculture Organisation showed sustainable increase throughout the years (1993–2005). They applied both the Malmquist index method and stochastic frontier translog production function methods. Though different methods presented different results for provincial (except Tibet) TFP growth, they provide valuable information about the difference in TFP growth among regions, showing that the Eastern region had higher TFP growth than the rest of China.

Using the Törnqvist index methodology, Wang *et al.* (2013) pointed out that the annual growth of agricultural TFP for selected 25 provinces in China was 2.7% between 1985 and 2007. They constructed a spatial-linked TFP index for 25 Chinese provinces and found that the faster TFP growth in coastal regions might imply a widening inequality between the coastal region and the rest.

Table 4.1: Summary of Selected Literature on Agricultural TFP in China

Authors	Study Period	Methodology	National Annual TFP Change (%)	Research Scope
Chen *et al.* (2008)	1990–2003	Malmquist Index, and Sequential Data Envelop Analysis (DEA)	1990–1996: 1.40; 1996–1999: 0.30; 1999–2003: 2.50	29 provinces
Fan and Zhang (2002)	1952–1997	Tornqvist–Theil Index	1952–1997: 1.82; 1979–1997: 3.28	National level (1952–1997); 25 Provinces (1979–1997)
Lambert and Parker (1998)	1979–1995	Divisia Multi-Factor Productivity Index	3.3	27 provinces, excluding Beijing, Tianjin and Shanghai
Li, Zeng and Zhang (2008)	1980–2005	Malmquist Index, DEA	2.2	30 provinces, excluding Tibet
Mao and Koo (1997)	1984–1993	Malmquist Index, DEA	1.42	29 provinces
Mead (2003)	1982–1999	Cobb–Douglas Production Function	1991: 0.81; 1993: −4.49; 1996: 15.59; 1997: 2.06; 1998: −0.47; 1999: −1.03	224 observations from 28 provinces, excluding Tibet and Hainan
Nin-Pratt, Yu and Fan (2010)	1961–2006	Malmquist Index, DEA	2.11	National level (total 60 countries)
Tong, Fulginiti and Sesmero (2009)	1993–2005	Malmquist Index, DEA; Stochastic Frontier Production Function	Malmquist Index, DEAs: 1.60; SPF: −0.18	29 provinces (excluded Tibet and Chongqing)
Wang and Rungsuriyawiboon (2010)	1991–2005	Frontier Function Approach	3.23	28 provinces, except for Hainan and Tibet
Wang *et al.* (2013)	1985–2007	Tornquist–Theil Index	2.7	25 contiguous provinces
Wu *et al.* (2001)	1980–1995	Malmquist Index, DEA	2.4	29 provinces before 1988 and 30 provinces from 1988 to 1995
Xu (1999)	1979–1996	Constant Elasticity of Substitution Production Function	2.3	National level

It is useful to note that the data on inputs and output vary among different studies. Output data can be divided into two categories broadly, *viz.* gross agricultural output value and agricultural value added. Mao and Koo (1997) suggested that since gross agricultural output included all inputs in agricultural output, the value-added agricultural output should be used to measure TFP. They calculated TFP growth using the Malmquist index method and found that the annual TFP growth for 29 provinces during 1984–1993 was 1.4%.

Chen *et al.* (2008) also adopted agricultural value added as output and applied the Malmquist index approach to analyse TFP growth in China for the period of 1990–2003. They found that annual TFP growth for 29 provinces was 1.5%. Their results indicated that the TFP growth declined from 1.4% in 1990–1996 to 0.3% in 1996–1999, and rose rapidly to 2.5% in 1999–2003. They concluded that the major driving factors for technological progress were the reduction in agricultural tax and the increase in public expenditure in R&D.

This chapter offers an up-to-date estimate of agricultural TFP growth in China at the national and provincial levels for the 1990–2012 period using alternative parametric and non-parametric methodologies. Understanding TFP growth in the 2000s in China is particularly relevant given that the central government consecutively issued "No. 1 Central Documents"[7] on agriculture sector for 11 years (2004–2014) with a focus on enhancing agricultural R&D and overall efficiency, as well as raising farmers' income.

4.3 METHODOLOGY AND DATA SOURCES

Broadly, TFP growth can be measured using two major approaches — parametric and non-parametric techniques. While more widely used, the accuracy of the parametric approach heavily depends on the specification of the production function form. An inaccurate specification of the production function could likely result in a misinterpretation of TFP. Parametric approaches also make strict assumptions of the functional form, such as constant-returns-to scale and technological efficiency, which are difficult to justify in practice. In contrast, the Malmquist index methodology is sensitive to outliers and data quality (Tong, Fulginiti and Sesmero 2009), but not dependant on model specification. This chapter applies growth accounting as the parametric approach and Malmquist index as the non-parametric approach to gain a comprehensive understanding of the agricultural TFP growth in China. Given a lack of observations to conduct panel analysis at provincial levels, we applied growth accounting methodology only at national level. In order to have a deeper understanding of the driving factors of TFP growth, we employ the Malmquist index method at the provincial level, which allows us to decompose TFP change into TC and TEC.

4.3.1 Growth Accounting Methodology

In this study, the production function for China's agriculture sector is defined in Cobb–Douglas production function form.

$$Y_{it} = A_{it} X_{1it}^{\beta_1} X_{2it}^{\beta_2} X_{3it}^{\beta_3}, \tag{4.3.1}$$

[7]"No. 1 Central Documents" refers to the first policy document of China's government for a year. It comes from the Central Committee of the Communist Party of China and the State Council. http://english.cntv.cn/20140120/100039.shtml (accessed January 26, 2015).

where i and t indicate economy and time, respectively. Y_{it} represents the agricultural value added per labour for economy i at time t. X_{1it}, X_{2it} and X_{3it} indicate machinery per labour, fertiliser per labour and land per labour for economy i at time t respectively. A_{it} is usually defined as the measure for TFP, which includes technology and efficiency factors.

By taking the natural logarithm on both sides of Equation (4.3.1), one can obtain Equation (4.3.2), which determines the contribution of inputs to output and is often used in calculating TFP growth.

$$\ln Y_{it} = \beta_0 + \beta_1 \ln X_{1it} + \beta_2 \ln X_{2it} + \beta_3 \ln X_{3it} + \varepsilon_{it}. \tag{4.3.2}$$

TFP growth is the difference between real agricultural output growth and the weighted sum of various inputs. Hence, TFP growth is calculated as the residual change in output. Therefore, Total Factor Productivity Change (TFPC) can be measured by Equation (4.3.3).[8]

$$TFPC_{it} = y_{it} - \hat{\beta}_1 x_{1it} - \hat{\beta}_2 x_{2it} - \hat{\beta}_3 x_{3it}. \tag{4.3.3}$$

4.3.2 Malmquist Index Data Envelopment Analysis

Considering various limitations of the parametric approach to estimating TFP, and given limited time series, this chapter also applied the non-parametric approach to measure agricultural TFP growth in China in the form of the Malmquist index.[9]

According to Färe *et al.* (1994), if a selected time point is represented by t, its corresponding technology level can be labelled as S^t. In this context, x^t can be used to represent all inputs of the production activity, and y^t can be used to represent all outputs of the production activity. Under these circumstances, S^t can be defined in the form of Equation (4.3.4) which captures all possible combinations of inputs x^t and outputs y^t under the available technology at time t.

$$S^t = \left\{(x^t, y^t), x^t \text{ produce } y\right\}^t. \tag{4.3.4}$$

Since agricultural production activity in the real world can hardly fully achieve and implement all cutting-edge technologies in a specific time point t, a differentiation between the actual output and the ideal output could be spotted. According to Färe *et al.* (1994), the differentiation can be defined as Equation (4.3.5).

$$D_y^t(\mathrm{x}^t, \mathrm{y}^t) = \inf\left\{\theta : (\mathrm{x}^t, \mathrm{y}^t/\theta) \in \mathrm{S}^t\right\} = \left(\sup\left\{\theta : (\mathrm{x}^t, \theta\, \mathrm{y}^t) \in \mathrm{S}^t\right\}\right)^{-1} \tag{4.3.5}$$

[8] $y_{it} = \ln Y_{it} - \ln Y_{i,t-1} \approx (Y_{it} - Y_{i,t-1})/Y_{i,t-1}$; $X_{jit} = \ln X_{jit} - \ln X_{ji,t-1} \approx (X_{jit} - X_{ji,t-1})/X_{ji,t-1}$ $(j = 1,2,3)$.

[9] The other commonly used non-parametric measure is the Törnqvist index. However this has rather strict data requirements, including prices and quantitative data of both inputs and output. In addition, it also has some assumptions such as a perfectly competitive market and profit maximisation, which are not suitable for agricultural studies.

Equation (4.3.5) is an output-oriented distance function, in which $D^t_y(x^t,y^t)$ represents the reciprocal of the maximum change in output y^t through fully exploiting the input x^t. Given the input level x^t at time t, Equation (4.3.5) can be used to measure the maximum expansion of the output y^t.

$$D^t_y(x^{t+1},y^{t+1}) = \inf\left\{\theta : (x^{t+1},y^{t+1}/\theta) \in S^t\right\} = \left(\sup\left\{\theta : (x^{t+1},\theta y^{t+1}) \in S\right\}\right)^{-1}. \quad (4.3.6)$$

Equation (4.3.6) represents the reciprocal of maximum change of output, given input x^{t+1}, to make y^{t+1} feasible at time t. The Malmquist index is defined in a distance function by Färe *et al.* (1994), as shown in Equation (4.3.7). In order to avoid choosing an arbitrage benchmark, the geometric mean of two Malmquist indices can be calculated to form a benchmark (Färe *et al.* 1994).

$$M_y(x^t,y^t,x^{t+1},y^{t+1}) = \left[\left(\frac{D^t_y(x^{t+1},y^{t+1})}{D^t_y(x^t,y^t)}\right)\left(\frac{D^{t+1}_y(x^{t+1},y^{t+1})}{D^{t+1}_y(x^t,y^t)}\right)\right]^{\frac{1}{2}} \quad (4.3.7)$$

By mathematical transformation, Equation (4.3.7) can be written in the form of Equation (4.3.8), which allows us to decompose the TFP change into TC and TEC (Färe *et al.* 1994).

$$M_y(x^t,y^t,x^{t+1},y^{t+1}) = \frac{D^{t+1}_y(x^{t+1},y^{t+1})}{D^t_y(x^t,y^t)}\left[\left(\frac{D^t_y(x^{t+1},y^{t+1})}{D^{t+1}_y(x^{t+1},y^{t+1})}\right)\left(\frac{D^t_y(x^t,y^t)}{D^{t+1}_y(x^t,y^t)}\right)\right]^{\frac{1}{2}} \quad (4.3.8)$$

As specified by Färe *et al.* (1994), Equation (4.3.8) can be written in a more illustrative form as Equation (4.3.9). TEC refers to activities which can enhance the efficiency of production, including better resource utilisation and management. TC refers to a set of developments that can move the productivity frontier forward (innovation, etc.). Linear Programming approach is used to calculate the Malmquist index in this chapter. In short, Malmquist index methodology not only allows us to decompose TFPC into TEC and TC, but also provides relative rankings of economies based on their distance to the efficiency frontier.

$$\text{TFPC} = \text{TEC} \times \text{TC} \quad (4.3.9)$$

4.3.3 Data Sources[10]

In agricultural statistics for China, there are two kinds of output value: one is gross agricultural production value and the other is agricultural value added. Gross agricultural production involves all the output value in the agriculture sector. Most studies used the gross agricultural output value, but it also includes all the inputs value of agricultural production (Mao and Koo 1997).

[10] Sichuan and Chongqing are combined in this chapter for the period of 1990–2012 as Chongqing was separated from Sichuan in 1997.

Therefore, this chapter uses agricultural value added as the output to measure TFP. Real Primary Sector GDP (Constant Year 1990 Prices, 100 Million Yuan) refers to the total value of products of agriculture, forestry, animal husbandry and fishery, and the total value of services in support of agriculture, forestry, animal husbandry and fishery activities, excluding intermediate costs. Data for output is derived from *China Statistical Yearbook*.

In terms of agricultural inputs, based on data availability, four major inputs are used in this chapter. First of all, labour is defined as the number of employed persons in the primary sector (10,000 persons). Data is taken from *Provincial Statistical Yearbooks* because data is missing for some provinces for some years in *China Statistical Yearbook*. Second, data on other inputs[11] such as land, fertiliser and machinery are collected from *China Rural Statistical Yearbook* (1991–2013). Land refers to sown area of crops (1,000 hectares), which is the area of land (cultivated or non-cultivated area) sown or transplanted with crops that are harvested within the calendar year by agricultural producers. Machinery refers to the total power of agricultural machinery (10,000 kW), which is the total capacity of all agricultural machinery used for planting, animal husbandry, fishery, primary processing of agricultural products, agricultural transport and infrastructure construction of farmland. Fertiliser refers to the consumption of chemical fertilisers in agriculture in the year (10,000 tons), including nitrogenous fertiliser, phosphate fertiliser, potash fertiliser and compound fertiliser. It is calculated as follows[12]:

$$V = Q \times E, \tag{4.3.10}$$

where

V = Volume of effective component;
Q = Physical quantity;
E = Effective component of certain chemical fertiliser (%).

4.4 RESEARCH FINDINGS

4.4.1 Labour Productivity

Before examining the results for TFP, Table 4.2 reports labour productivity in China, defined as the ratio of the volume measure of output divided by the measure of input use.[13] Around half of the 30 provinces had higher than average annual growth in labour productivity. Jiangsu had the highest average annual growth of labour productivity in mainland China from 1991 to 2012. Note that provinces in Western China, such as Sichuan and Chongqing, Shaanxi, Qinghai and Gansu had labour productivity growth above the national average. However, Tianjin, Guangdong, Shanghai and Beijing ranked bottom in terms of labour productivity growth. Comparing the two columns in Table 4.2, one can observe that some provinces such as Hebei and Hainan ranked top in both agricultural output and labour productivity growth. However, these are exceptions as

[11]The detailed definitions for other inputs used in this chapter are derived from *China Statistical Yearbook 2013*.

[12]Equation (14) is provided by *China Statistical Yearbook, Statistics Description*.

[13]This definition of labour productivity is derived from Freeman (2008).

Table 4.2: Annual Growth of Real Output and Labour Productivity (1991–2012)

Province	Output Growth (%)	% Change in Labour Productivity
Jiangsu	5.05	9.42
Zhejiang	3.69	8.42
Hebei	6.94	8.25
Sichuan and Chongqing	5.32	7.72
Shaanxi	6.35	7.60
Hainan	7.36	7.02
Hunan	5.63	6.99
Fujian	5.86	6.89
Guangxi	6.36	6.80
Anhui	5.54	6.71
Qinghai	5.37	6.60
Henan	6.01	6.53
Jilin	6.25	6.42
Gansu	6.36	6.34
Jiangxi	4.52	6.21
Hubei	5.37	6.04
Liaoning	6.32	6.00
Inner Mongolia	6.44	5.86
Shandong	4.98	5.80
Heilongjiang	6.97	5.68
Xinjiang	6.92	5.65
Shanxi	5.60	5.53
Guizhou	4.66	5.43
Yunnan	5.35	5.09
Ningxia	5.98	4.91
Tianjin	3.15	4.45
Guangdong	3.62	4.39
Shanghai	0.95	4.30
Beijing	0.22	2.42
Tibet	2.04	1.69
Mean	**5.17**	**6.04**

Source: China Statistical Yearbook, Provincial Statistical Yearbooks

most provinces only performed well in one of the indicators. Heilongjiang, for instance, ranked second in agricultural output growth rate, while ranking 20th in terms of labour productivity growth.

4.4.2 Total Factor Productivity

As is well known, labour productivity growth could arise either due to TFP growth or due to increases in non-labour inputs, the latter being costly and subject to diminishing returns. As shown in Tables 4.3 and 4.4, the average TFP growth for agriculture in China based on the Cobb–Douglas production function and Malmquist index during the sample period was 2.4% and 2.8%, respectively. This narrow range of estimates suggests a degree of robustness in the estimates at the national level.

In addition, results from Malmquist index methodology suggest that TFP growth can be predominantly attributed to TC, with the growth in TEC being negative for 20 out of 30 provinces. One can observe significant variations in TFP growth rate across the provinces, primarily driven by TC. TC was larger than 1 for all provinces in China with an average growth rate of 3.1%, except for Tibet and Qinghai. This finding indicates that China's agricultural innovation and technological progress has been significant in the past 20 years. Meanwhile, TEC was below 1 for most of the provinces, implying that agricultural efficiency has been decreasing in the past two decades.

Of the top 10 provinces for labour productivity growth listed in Table 4.2, four also show up in the top 10 for TFP growth (Jiangsu, Hebei, Fujian and Zhejiang). Sichuan and Chongqing, Hunan and Shaanxi which performed well for labour productivity growth, moved out of the top 10 for TFP growth. Liaoning, Shandong, Xinjiang and Shanghai performed below average for labour productivity growth but moved into the top 10 provinces with respect to TFP growth. Jiangsu, Hebei and Zhejiang stood out, being consistently ranked in the top five for both labour productivity and TFP growth. Their TFP growth was almost double to that of the national mean due to their above average performance for both TC and TEC.

When these provinces are grouped into regions, TFP growth varied significantly. The North-eastern provinces (Heilongjiang, Jilin and Liaoning), often considered as the food baskets of China, rank moderately for labour productivity. Meanwhile, Jilin and Liaoning were ranked among the top 10 provinces for TFP growth. In the Western part of China, Shaanxi consistently performed well, ranking among the top five for labour productivity growth and the top 15 for TFP growth. Xinjiang, whose labour productivity growth was below average,

Table 4.3: Cobb–Douglas Annual Average TFP Growth

	β_0	β_1	β_2	β_3	TFPC
National Average	0.021***	0.397***	0.137***	0.439***	2.4%
	(5.52)	(7.85)	(2.93)	(5.16)	

Note: *t*-statistics are reported in parentheses
*, **, and *** indicate statistically significant at the 10%, 5% and 1% levels, respectively
Source: Authors' calculation

Table 4.4: Malmquist Index, Summary of Provincial Means (1990–2012)

Rank	Province	TEC	TC	TFPC
1	Jiangsu	1.021	1.052	1.074
2	Hebei	1.019	1.038	1.057
3	Fujian	1.005	1.051	1.056
4	Zhejiang	1.014	1.038	1.053
5	Jilin	0.995	1.056	1.051
6	Liaoning	1.002	1.045	1.048
7	Shandong	0.996	1.048	1.044
8	Xinjiang	0.991	1.053	1.044
9	Henan	0.993	1.05	1.042
10	Shanghai	0.996	1.045	1.041
11	Anhui	0.989	1.048	1.036
12	Shaanxi	1.005	1.03	1.035
13	Heilongjiang	0.994	1.04	1.033
14	Hainan	1	1.033	1.033
15	Guangdong	0.988	1.041	1.029
16	Shanxi	0.989	1.039	1.027
17	Inner Mongolia	0.994	1.033	1.027
18	Ningxia	0.987	1.04	1.027
19	Qinghai	1.026	1	1.026
20	Beijing	0.991	1.033	1.024
21	Gansu	1.014	1.005	1.02
22	Hunan	1.004	1.014	1.017
23	Jiangxi	0.991	1.024	1.014
24	Tianjin	0.987	1.026	1.012
25	Hubei	0.987	1.025	1.012
26	Guangxi	0.994	1.018	1.012
27	Sichuan and Chongqing	0.99	1.02	1.01
28	Yunnan	0.983	1.007	0.99
29	Guizhou	0.972	1.016	0.988
30	Tibet	0.993	0.977	0.97
	Mean	**0.997**	**1.031**	**1.028**

Source: Authors' calculation

Table 4.5: Results Excluding the Top-3 and Bottom-3 Provinces

Rank	Province	TEC	TC	TFPC
1	Jilin	0.999	1.053	1.051
2	Liaoning	1.003	1.045	1.048
3	Zhejiang	1.009	1.039	1.048
4	Shandong	0.997	1.049	1.046
5	Henan	0.994	1.05	1.044
6	Xinjiang	0.994	1.05	1.044
7	Shanghai	0.997	1.044	1.041
8	Anhui	0.99	1.047	1.037
9	Shaanxi	1.006	1.029	1.036
10	Heilongjiang	0.997	1.037	1.034
11	Hainan	1	1.03	1.03
12	Guangdong	0.988	1.041	1.028
	Mean of top 12	**0.998**	**1.043**	**1.041**
13	Shanxi	0.989	1.038	1.027
14	Ningxia	0.989	1.039	1.027
15	Beijing	0.991	1.033	1.024
16	Qinghai	1.006	1.018	1.024
17	Inner Mongolia	0.994	1.03	1.023
18	Hunan	1.004	1.015	1.019
19	Jiangxi	0.991	1.024	1.014
20	Guangxi	0.994	1.019	1.013
21	Gansu	1.002	1.011	1.013
22	Hubei	0.988	1.024	1.012
23	Tianjin	0.987	1.024	1.011
24	Sichuan and Chongqing	0.99	1.021	1.011
	Mean of bottom 12	**0.994**	**1.025**	**1.018**

Source: Authors' calculation

ranked in the top 10 for TFP growth. Tibet remained a laggard province in terms of both labour productivity and TFP.

Excluding the Top-3 and Bottom-3 provinces (outliers) and re-estimating the Malmquist index, we found that the ranking remained similar (Table 4.5). The variation between the top half and the bottom half was driven by TC. Three North-eastern provinces — Heilongjiang, Jilin and Liaoning — remained in the top 10. Most Western provinces were lagging in terms of TFP growth, except Xinjiang province. Comparing the provinces in the top half against those in the

Table 4.6: National Average of TEC, TC and TFPC in Sub-Periods

	TEC	TC	TFPC
1990–2012	0.997	1.031	1.028
1990–2000	0.977	1.029	1.006
2001–2012	1.014	1.033	1.047

Source: Authors' calculation

bottom half, one finds that the primary difference lies in TC, which is related to agricultural technology and innovation.

When the whole period was divided into two sub-periods, results were similar to those of Chen *et al.* (2008). As shown in Table 4.6, TFP growth was 0.6% in the 1990s and increased significantly to 4.7% from 2001 to 2012. Both TEC and TC increased rapidly during the second period, implying that the government policy to modernise agriculture has been effective (Jiang and Cui 2011). Since 2004, every year, the central government of China has consecutively issued "No.1 Central Documents" that emphasise on enhancing agricultural productivity and rural development.

In 2004, China started to reduce agricultural tax and at the same time subsidise farmers for cultivating staple food crops, and encourage purchasing of high-quality seeds and agricultural machinery. As shown in Table 4.7, there was an increase in TFPC, especially in TEC. In 2005, a policy was issued to enhance the production capacity of the agriculture sector through increases in financial, governmental and technical support. The policy issued in 2006[14] and 2007 focussed on deepening rural reforms and developing modern agriculture. During 2005–2007, China maintained moderate TFP growth as shown in Table 4.7. In 2008, a policy was issued to enhance agricultural infrastructure and increase farmers' income. One can find in Table 4.7 that the TEC and TC increased 1.4% and 5.7%, respectively in this year.

In response to the global food crisis in 2007–2008, the central government urged authorities to ensure steady development of agricultural capacity and increase farmers' income in 2009 and 2010. Attention has been focused on rural infrastructure and innovation in 2011 and 2012, when policies to enhance rural infrastructure and accelerate agricultural innovation were introduced. One can observe from Table 4.7 that annual TFP increased by around 6.7% in 2011 and 2012.[15] Overall, the TFP growth between 2004 and 2012 was broadly higher than that before 2004, which indicates that the agricultural reforms initiated by the central government had a significant impact on China's agricultural productivity (Mead 2003).

Based on the empirical results, Jiangsu and Hebei performed exceedingly well for both labour productivity and TFP growth, as listed in Tables 4.2 and 4.4. In addition to persistent investments in agricultural R&D and policies by their local governments promoting agricultural innovation, their location (coastal area) and fertile land endowments also contributed significantly to the

[14]Agricultural tax was exempted throughout the country in 2006.

[15]http://english.cntv.cn/20140120/100039.shtml (accessed January 26, 2015).

Table 4.7: National Average of TEC, TC and TFPC 2001–2012

Year	TEC	TC	TFPC
2001	1.01	1.002	1.012
2002	0.954	1.088	1.038
2003	1.072	0.963	1.033
2004	1.107	0.974	1.078
2005	1.005	1.019	1.024
2006	0.988	1.043	1.03
2007	1.069	0.971	1.038
2008	1.014	1.057	1.072
2009	0.997	1.051	1.048
2010	0.951	1.118	1.063
2011	0.986	1.083	1.068
2012	1.025	1.041	1.067
Mean	1.014	1.033	1.047

Source: Authors' calculation

high productivity through improving technical efficiency. In contrast, geographical conditions turned out to be constraints for Yunnan and Tibet, with their special climate and geographical conditions (plateau and mountainous areas) preventing them from sharing the technical skills with provinces in plain areas, hindering their efficiency improvements. Further, they also suffered from lacking R&D facilities and investments, which limited their agricultural technological progress.

4.5 CONCLUDING REMARKS AND POLICY IMPLICATIONS

With a population of 1.37 billion[16] and limited resources, enhancing agricultural productivity is paramount for China. In recent years, more agricultural land has been converted to industrial land and a significant proportion of the labour force previously engaged in the agriculture sector has shifted to the secondary and tertiary industries in cities and towns. Felipe, Dacuycuy and Lanzafame (2014) projected that the share of employment in the primary sector will decline to 24% by 2020, which means that China needs to enhance its agricultural TFP to remain self-sufficient.

The empirical results in this chapter provide up-to-date results of agricultural TFP for mainland China. During the sample period of 1990–2012 there was fairly rapid growth in TFP (2.4–2.8%) in China, especially in the second period from 2001 to 2012 (4.7%). Technical

[16]Please refer to Footnote 1 of this chapter regarding the size of the population in mainland China.

Figure 4.7: Public Agricultural R&D Spending in Brazil, China, India, Malaysia, Mexico and Nigeria 1991–2008

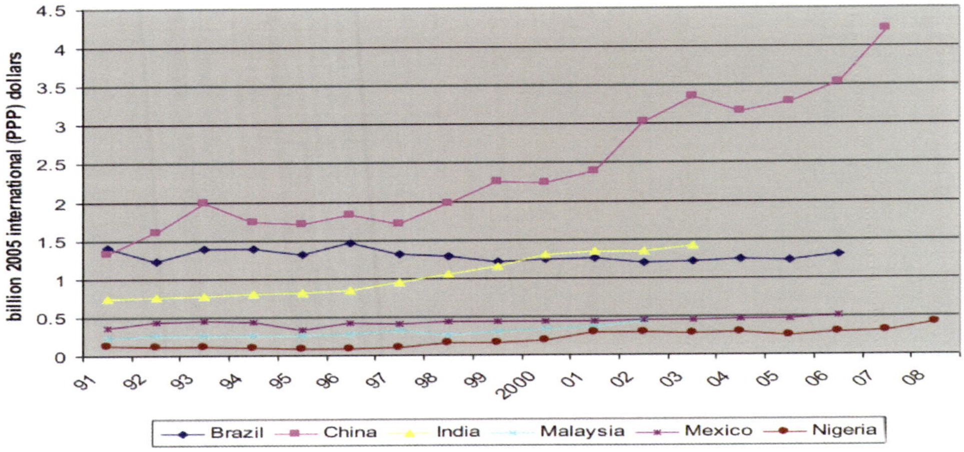

Source: Sharma and Gulati (2012)

change (TC) has generally experienced an improving trend for most provinces, except Tibet. This is consistent with the emphasis on agricultural R&D spending as well as the use of high-yield seeds. As shown in Figure 4.7, China experienced much faster growth in public agricultural R&D spending than other developing countries in 1991–2008, especially after 2000 when we can see a clear divergent trend, whereby China further entrenched its lead in areas key to agricultural productivity enhancement.

Since the mid-1980s, the agricultural R&D funding system in China has been decentralised (Fan, Qian and Zhang 2006). Local research institutes are encouraged to fund themselves and many are becoming more commercialised. While it helps to link agricultural research to farms and push researchers to be more productive, which in turn is reflected in improved TC performance, there are valid concerns that such a decentralised approach to R&D spending leads to overlap in spending and fails to benefit from scale economies that a more coordinated approach across provinces might provide.[17]

In addition, of concern is the fact that TEC is generally stagnant across provinces with some exceptions. The experience of successful provinces such as Jiangsu and Hebei may provide some useful lessons to other provinces in China and offer opportunities for inter-province learning to improve TEC. First, improvement in irrigation and agricultural infrastructure is paramount to enhance productivity. Second, there is a need to enhance the collaboration between research institutes (including tertiary education institutions) and markets, and to build well-known agricultural product brands that can significantly increase the value added in the entire value chain involving the agricultural industry. Third, local governments need to take the initiative to assist

[17] In contrast, India's approach to agricultural R&D spending is too centralized and bureaucratic and fails to appreciate important differences between states (Pal and Byerlee 2006).

farmers to collaborate with companies. This will enable the farmers to extend the production line, thus raising farm wages and connecting agricultural production closely with market demand.

Those provinces that are lagging in TEC can learn from the well-performing provinces in terms of expanding agricultural mechanisation, improving extension services and establishing efficient production value chain. It is worth noting that different provinces have different natural conditions, which will affect provincial TEC or TC. Therefore, when new technologies or technical skills are promoted at the provincial level, it is important to localise technologies to match local demand conditions.

REFERENCES

Chen, Po-Chi, Ming-Miin Yu, Ching-Cheng Chang and Shih-Hsun Hsu. 2008. "Total Factor Productivity Growth in China's Agricultural Sector." *China Economic Review* 19: 580–593.

DeBrauw, A., J. Huang and S. Rozelle. 2004. "The Sequencing of Reforms in China's Agricultural Transition." *Economics of Transition* 12(3): 427–466.

Deng, Hengshan, Jikun Huang, Zhigang Xu and Scott Rozelle. 2010. "Policy Support and Emerging Farmer Professional Cooperatives in Rural China." *China Economic Review* 21(4), 495–507.

Fan, S. and P. Pardey. 1997. "Research Productivity and Output Growth in Chinese Agriculture." *Journal of Development Economics* 53(1): 115–137.

Fan, Shenggen and Xiaobo Zhang. 2002. "Production and Productivity Growth in Chinese Agriculture: New National and Regional Measures." *Economic Development and Cultural Change* 50(4): 819–838.

Fan, Shenggen, Keming Qian and Xiaobo Zhang. 2006. "China: An Unfinished Reform Agenda." In *Agricultural R&D in the Developing World: Too Little, Too Late?*, eds. P.G. Pardey, J.M. Alston and R.R. Piggott, 29–63. Washington, DC: International Food Policy Research Institute.

Färe, Rolf, Shawna Grosskopf, Mary Norris and Zhongyang Zhang. 1994. "Productivity Growth, Technical Progress, and Efficiency Change in Industrialized Countries." *The American Economic Review* 84(1): 66–83.

Felípe, Jesus, Connie Bayudan-Dacuycuy and Matteo Lanzafame. 2014. "The Declining Share of Agricultural Employment in the People's Republic of China: How Fast?" Asian Development Bank Economics Working Paper Series 419.

Freeman, Rebecca. 2008. *Labour Productivity Indicators.* Paris: Organisation for Economic Co-operation and Development (OECD).

Huang, Jikun and Scott Rozelle. 2015. "China's Agriculture: Past Failures, Present Successes and Enabling Policies." In *Routledge Handbook of the Chinese Economy*, eds. G.C. Chow and D.H. Perkins, 168–181. New York: Routledge.

Jiang, Heping and Qifeng Cui. 2011. "Experience of Modern Agricultural Development in China." http://www.unctad.info/upload/GCF2011/doc/A4-A8/gcf2011_A8_Qifeng_en.pdf.

Lambert, David K. and Elliott Parker. 1998. "Productivity in Chinese Provincial Agriculture." *Journal of Agricultural Economics* 49(3): 378–392.

Lardy, N.R. 1983. *Agriculture in China's Modern Economic Development.* Cambridge: Cambridge University.

Li, G., X. Zeng and L. Zhang. 2008. "Study of Agricultural Productivity and Its Convergence across China's Regions." *The Review of Regional Studies* 38(3): 361–379.

Lin, J. 1992. "Rural Reforms and Agricultural Growth in China." *American Economic Review* 82(1): 34–51.

Mao, Weining and Won W. Koo. 1997. "Productivity Growth, Technological Progress, and Efficiency Change in Chinese Agriculture after Rural Economic Reforms: A DEA Approach." *China Economic Review* 8(2): 157–174.

Mead, Robert W. 2003. "A Revisionist View of Chinese Agricultural Productivity?" *Contemporary Economic Policy* 21(1): 117–131.

National Bureau of Statistics of China. 1981–2012. *Statistical Yearbook of China, various issues from 1981–2012.* Beijing: China Statistical Press.

Nin-Pratt, Alejandro, Bingxin Yu and Shenggen Fan. 2010. "Comparisons of agricultural productivity growth in China and India." *Journal of Productivity Analysis* 33: 209–223.

Pal, Suresh, and Derek Byerlee. 2006. "India: The Funding and Organization of Agricultural R&D — Evolution and Emerging Policy Issues." In *Agricultural R&D in the Developing World: Too Lttle, Too Late?*, 155.

Sharma, Pooja and Ashok Gulati. 2012. *Approaches to Food Security in Brazil, China, India, Malaysia, Mexico and Nigeria: Lessons for Developing Countries.* ICRIER Policy Series No. 14.

Studwell, Joe. 2014. *How Asia Works: Success and Failure in the World's Most Dynamic Region.* New York: Grove Press.

Tan, Khee Giap, Haoran Suo and Ramkishen S. Rajan. "Estimating Provincial Agricultural Total Factor Productivity in Mainland China." *International Journal of Business Performance Management*, Forthcoming.

Tan, Khee Giap, Randong Yuan and Wei Cher Sangiita Yoong. 2015. *2014 Annual Provincial and Regional Competitiveness Analysis for Greater China Economies: Development Strategies under the New Normal.* Singapore: World Scientific Publishing Company.

Tong, Haizhi, Lilyan E. Fulginiti and Juan P. Sesmero. 2009. *Chinese Regional Agricultural Productivity: 1994–2005.* Lilyan E. Fulginiti Publications, Paper 10.

Wang, Sun Ling, Francis Tuan, Fred Gale, Agapi Somwaru and James Hansen. 2013. "China's regional agricultural productivity growth 1985-2007: A multilateral comparison." *Agricultural Economies* 44(2): 241–251.

Wang, Xiaobing and Supawat Rungsuriyawiboon. 2010. "Agricultural Efficiency, Technical Change and Productivity in China." *Post-Communist Economies* 22(2): 207–227.

Wen, G. 1993. "Total Factor Productivity Change in China's Farming Sector: 1952–1989." *Economic Development and Cultural Change* 42(1): 1–41.

Wu, Shunxiang, David Walker, Stephen Devadoss and Yao-chi Lu. 2001. "Productivity Growth and its Components in Chinese Agriculture after Reforms." *Review of Development Economics* 5(3): 375–391.

Xu, Yingfeng. 1999. "Agricultural Productivity in China." *China Economic Review* 10(2): 108–121.

APPENDIX: THE MALMQUIST INDEX DATA ENVELOPMENT ANALYSIS (DEA) METHODOLOGY[18]

According to Färe *et al.* (1994), the change of TFP can be decomposed into TC and TEC through Malmquist index. Every time point has its specific technology level. Therefore, if a selected time point is represented by t, its corresponding technology level can be labelled as S^t. In this context, x^t can be used to represent all inputs of the production activity, and Y^t can be used to represent all outputs of the production activity. Under these circumstances, S^t can be defined in the form of Equation (A.1) which captures all possible combinations of inputs x^t and outputs y^t under the available technology at time t.

$$S^t = \left\{(x^t, y^t), x^t \text{ produce } y\right\}^t. \tag{A.1}$$

Since the agricultural production activity in the real world can hardly fully achieve and implement all cutting-edge technologies in a specific time point t, a differentiation between the actual output and the ideal output could be spotted. According to Färe *et al.* (1994), the differentiation can be defined as Equation (A.2).

$$D_y^t = \inf\left\{\theta : (\mathrm{x}^t, \mathrm{y}^t/\theta) \in \mathrm{S}^t\right\} = \left\{\sup\left[\theta : (\mathrm{x}^t, \theta\, \mathrm{y}^t) \in \mathrm{S}^t\right]\right\}^{-1} \tag{A.2}$$

Equation (A.2) is an output-oriented distance function, in which D_y^t represents the reciprocal of the maximum change in output y through fully exploiting the input x. Given the input level x^t at time t, Equation (A.2) can be used to measure the maximum expansion θ of the output y^t. Meanwhile, D_y^t can also be interpreted as the reciprocal of the maximum efficiency change of the output y^t by fully exploiting input x^t.

Similarly, as shown in Equation (A.3), the input-oriented distance function can measure the change of inputs given the output level.

$$D_x^t = \sup\left\{\lambda : (x^t / \lambda, y^t) \in S^t\right\} \tag{A.3}$$

$$D_y^t = \inf\left\{\theta : (x^{t+1}, y^{t+1} / \theta) \in S^t\right\} = \left\{\sup[\theta : (x^{t+1}, \theta y^{t+1}) \in S^t\right\}^{-1} \tag{A.4}$$

Equation (A.4) represents the reciprocal of maximum change of output, given input x^{t+1}, to make y^t feasible at time t. Hence, Malmquist index can be written in the form of Equations (A.5) and (A.6).

$$M_y^t = \frac{D_y^t(x^{t+1}, y^{t+1})}{D_y^t(x^t, y^t)} \tag{A.5}$$

$$M_y^{t+1} = \frac{D_y^{t+1}(x^{t+1}, y^{t+1})}{D_y^{t+1}(x^t, y^t)} \tag{A.6}$$

[18]For details on Malmquist index method please refer to ACI Technical Working Paper No. 16.

If a specific time t is selected, the corresponding technology level will become the reference technology level. Under these circumstances, Equation (A.5) can be used to measure the necessary change to generate the output y^{t+1} through utilising the input x^{t+1} with the feasible technology S^t. In order to avoid choosing an arbitrage benchmark, the geometric mean of two Malmquist indices can be calculated to form a benchmark (Färe *et al.* 1994).

$$M_y(x^t,x^{t+1},y^t,y^{t+1})=\sqrt{M_y^t * M_y^{t+1}} = \left[\left(\frac{D_y^t(x^{t+1},y^{t+1})}{D_y^t(x^t,y^t)}\right)\left(\frac{D_y^{t+1}(x^{t+1},y^{t+1})}{D_y^{t+1}(x^t,y^t)}\right)\right]^{\frac{1}{2}} \quad (A.7)$$

Equation (A.7) can be decomposed into two parts through mathematical transformation as illustrated by Equation (A.8).

$$M_y(x^t,x^{t+1},y^t,y^{t+1})=\frac{D_y^{t+1}(x^{t+1},y^{t+1})}{D_y^t(x^t,y^t)}\left[\left(\frac{D_y^t(x^{t+1},y^{t+1})}{D_y^{t+1}(x^{t+1},y^{t+1})}\right)\left(\frac{D_y^t(x^t,y^t)}{D_y^{t+1}(x^t,y^t)}\right)\right]^{\frac{1}{2}} \quad (A.8)$$

Equation (A.8) can be perceived as two components. The first component represents the TEC between time t and time t+1, and the second component represents the TC between time t and time t+1. The practical meaning of this transformation is that, the TFPC can be decomposed into TEC and TC, as shown in Equation (A.9) (Färe *et al.* 1994).

$$\text{TFPC} = \text{TEC} \times \text{TC} \quad (A.9)$$

TEC is broadly related to better resource utilization or deployment as well as more efficient management. TC generally refers to the technology innovation or progress, which can shift the production frontier. For example, an invention of new technology could shift the agricultural production frontier. Thereafter, the true driving factor could be identified through the decomposition of TFPC, which may potentially draw valuable lessons for future development of the agriculture sector.[19]

Linear Programming approach is used to calculate the Malmquist index in this chapter. In short, Malmquist index methodology not only allows us to decompose TFPC into TEC and TC, but also provides relative rankings of economies based on their distance to the efficiency frontier.

[19]TEC can be further decomposed into two parts which are Pure Technical Efficiency Change and Scale Efficiency Change as illustrated by Equation (A.10).

$$M_y(x^t,x^{t+1},y^t,y^{t+1})=\frac{D_y^{v,t+1}(x^{t+1},y^{t+1})}{D_y^{v,t}(x^t,y^t)}\left[\left(\frac{D_y^{v,t}(x^t,y^t)}{D_y^{v,t+1}\left(x^{t+1},y^{t+1}\right)}\right)\left(\frac{D_y^{c,t+1}(x^{t+1},y^{t+1})}{D_y^{c,t}(x^t,y^t)}\right)\right]\left[\left(\frac{D_y^{c,t}(x^{t+1},y^{t+1})}{D_y^{c,t+1}(x^{t+1},y^{t+1})}\right)\left(\frac{D_y^{c,t}(x^t,y^t)}{D_y^{c,t+1}(x^t,y^t)}\right)\right]^{\frac{1}{2}}$$

(A.10)

Discussant Notes 1

A Review of the "Agricultural Productivity in Greater China Economies: Estimating Provincial Agricultural Total Factor Productivity in Mainland China"

Dr Zhang Qingjin
Director, Institute of Rural Development
Shandong Academy of Social Sciences
People's Republic of China

At the outset, I would like to say that I truly agree with this report. It contains many conclusions that are very much in line with the state of agricultural development in China. For instance, the report states that agriculture in China can help China's rural residents come out of poverty and increase income, while advances in technology help to promote agricultural growth. Investment and subsidies by the government also serve to drive agricultural development. All of these are relatively consistent with China's current situation. The latest development of agriculture in China verifies the conclusions of this report and there have been clear improvements in land and labour productivity in agriculture in China. However, increases in labour productivity far exceed the increases in land productivity and the contribution of the total factor productivity (TFP) to increasing agricultural productivity has steadily augmented. Now, I would like to give my own views on a number of issues that this report touches on.

First, there is a great disparity in the income from planting grain crops and cash crops in Chinese agriculture. This will have a considerable impact on calculations of the productivity of agricultural labour. Typical planting practices in Shandong are to plant one season of wheat and one season of corn. In conducting surveys in Shandong, we found that this method of rotating wheat and corn usually produces an annual income of around 1,000 Renminbi. However, the difference between this and cash crops is very large. In 2011, the price of apples was very high and surveys in Yantai and Weihai showed that the average income per mu of land was 30,000 Renminbi. From this perspective, the large difference in labour productivity for grain crops and cash crops is obvious. Currently, land transfer costs in Shandong have reached somewhere around 1,000 Renminbi per mu and the large amount of commercial capital that has come to agriculture means a large volume of land transfers. If they continue to plant grain crops, revenue generated will barely reach the cost of the land, which will obviously be a bad deal for investors. This is why many commercial investors choose to plant cash crops with high value added to ensure profits.

One of China's basic national strategies is grain self-sufficiency, which means that there is a great deal of disagreement regarding how much grain each province should have to plant. Shandong, Heilongjiang and Henan are major grain producing provinces and the central government places great pressure on these provinces to produce grain, requiring them to maintain a certain land area that is used to plant grain. This is not conducive to increasing the income of farmers, creating a source of conflict. In this report, none of the biggest grain producing provinces (Henan, Shandong and Heilongjiang) are ranked at the top for TFP. Could this be because the excessive use of land in grain production has not been accounted for? Compared with the rest of the country, their agricultural production methods are objectively very advanced and their efforts contributed greatly to the advancements in agricultural technology and planting methods.

Second, I would like to discuss the effect of the phenomenon of part-time employment on labour productivity, which is linked to labour input. Today most of China's agricultural production is carried out by rural households, but many of these households also engage in part-time work, working in the fields during the busiest times of the year and finding work in other areas rest of the time. Based on the current degree of mechanisation, our surveys show that labour input required to plant 1 mu of wheat is 2.5–3 days. Clearly, there is very little labour input needed at the current level of mechanisation. There is an even more extreme example of this kind in Qingdao where the high level of industrialisation provides the rural residents there with local employment opportunities, which means that farmers can return to planting after they finish work, which means that land is not transferred. However, if a farmer were to go to Guangdong for work, he would likely transfer his land. This demonstrates how part-time labour can greatly impact our calculation of labour input, which is worth taking into consideration.

Third, there are several factors in TFP that are related to sustainable development. Increases in mechanisation are clearly beneficial to development, but the use of fertiliser and the fertility of the soil can also impact the sustainability of agricultural development in China. The quality of the agricultural labour force in China is declining rapidly with only the elderly and the children who are left behind to tend the fields. Excessive use of fertilisers is also a major issue facing agricultural development in China. To remain agriculturally self-sufficient, a relatively high level of production per unit of land must be maintained, which requires the use of fertiliser. The quality of land is also in rapid decline. There are two reasons for this. One reason is the use of chemical pesticides, which has led to a drop in the concentration of organic compounds in soil. The other reason is the urbanisation, which in many areas has taken over good land, leaving poor quality soil to be farmed. When conducting TFP analysis, we should note that there are several factors that have different impacts on sustainable development in China.

Fourth, it is about the issue of agricultural subsidies. The conclusion in this report is correct, that agricultural subsidies most definitely stimulate agricultural development, but the impact on grain is not very large. We have carried out many surveys in Shandong where grain subsidies are given directly to farmers, but the administrative costs of doing this are very high. Rural government workers are very frustrated by this and often complain about how much trouble it creates for them. Another problem here is that direct grain subsidies are given based on the area planted and their amounts are calculated by assessing how much land has been planted, which requires a great deal of work. In a village of 500 households, it is nearly impossible to calculate the area planted for each household because of the huge amount of work involved. This means that many government workers provide direct subsidies based on the amount of land farmed, which is unrelated to the total area planted with grain, which means that subsidies are given even if vegetables or trees are planted on the land in question. This creates a considerable problem for direct grain subsidies, but as this is a central government policy and local governments do not have the authority to change it, it creates a great deal of administrative problems for them. For instance, some local governments have to provide water to farmers and want to collect water fees, but the farmers do not pay it, causing some village committees to suggest withholding a portion of direct grain subsidies to ensure they can collect the water fees. But this is actually not an option for the local governments as the central government has again insisted that direct grain subsidies must not be withheld and must be put directly in the hands of farmers. Furthermore, this subsidy has not been effective in increasing grain production. This is a topic that must be examined and discussed.

Chapter 5

Conclusion and the Way Forward to an Innovation Driven Economy for Greater China

5.1 THE DEMAND SIDE CHALLENGES TO THE CHINESE ECONOMY

After more than three decades of steadily rapid economic growth that contributed to China's phenomenal socioeconomic transformation,[1] the growth momentum of the Chinese economy has shown signs of slowing down. While the global economy has been undergoing some deep structural changes with grave implications that are still unfolding since the aftermath of the 2008 Global Financial Crisis (GFC), the Chinese economy has been growing at a noticeably slower speed in the last six years, from 10.4% in 2010 to under 8% since 2012, at 7.4% in 2014 and at 6.9% in 2015, the slowest since 1990,[2] indicating a clear downward trend that makes one wonder where the prospect of the further economic development of China lies.

Being the second largest economy and the largest trading nation in the world, China has become an increasingly important engine of growth to many other countries (Jenkins, Peters and Moreira 2008; Tan, Abeysinghe and Tan 2015). Consequently, whether China can maintain a stable level of growth momentum to further develop its economy is not only an issue that is critical for the benefits of its 1.37 billion population,[3] but it also carries significant influence on many other countries with China being one of their most important trade and investment partners. Therefore, how the Chinese economy will fare in the coming years and decades matters for the stable and sustainable development of the region and the world at large.

It has been widely acknowledged that China has now entered into the New Normal Era, in which the country is expected to achieve slower but sustainable, quality growth for several

---

[1]Since the reform and opening up of China in the late 1970s, China has enjoyed over three decades of stable and fast growth of its Gross Domestic Product (GDP) at an impressive average annual rate of 9.8% from 1978 to 2013. The annual average GDP growth rate for China for the period from 1978 to 2013 is calculated using data from the World Bank.

[2]China's GDP growth rate was 7.4% in 2014 and 6.9% in 2015 according to figures released by the National Bureau of Statistics of China. China's GDP growth rates from 1961 until 2013 were available from the World Bank.

[3]At the time of writing, the population of mainland China ticked at 1,374,539,658, according to the China's population clock provided by the National Bureau of Statistics of China website, accessible at http://data.stats.gov.cn/.

decades to come, according to President Xi Jinping and Premier Li Keqiang.[4] However, if such deceleration in economic growth continues in the next five to 10 years, to levels under 5% or even lower, China will face a tangible risk of falling into the middle-income trap where the "New Normal" could become "Abnormal"[5] and stagnation in per capita income may derail further economic development, which Mr Lou Jiwei, the Finance Minister of China, sees a greater than 50% chance of happening.[6] With a rapidly ageing population (Cai 2010; Galor 2012) and an urbanising economy that is in many ways still underdeveloped (Tao and Xu 2007), there are serious challenges China faces and has to overcome if it wants to maintain a steadfast pace of development in order to avoid falling into the middle-income trap and eventually graduate into a socioeconomically advanced country.

These challenges are lurking in many corners of the economy, and we can find their pronounced presence from the demand side of the economy, ranging from government expenditures in infrastructure development to the troika of investment, consumption and exports, which are recognised as the traditional drivers of the growth of the Chinese economy. We are already witnessing a fall in foreign direct investment (FDI), due to diminishing demographic dividends (Zhang, Yang and Wang 2011; Cai 2010), which will cause labour cost to rise in China, making the country increasingly less appealing to FDI in labour-intensive industries. The room for further infrastructure investment inside China may be increasingly limited as the build-up for basic industries is near saturation, which will serve to dampen the demand for non-FDI financed investment as government-led infrastructure development is a major component of such investment.

With domestic demand contributing to less than 40% of GDP in China in the last decade,[7] household consumption is still not playing the role of the leading driver of growth, and it may not be able to assume that role in the short term either, largely due to the lack of a well-functioning social security in China to provide sufficient protection for the poor, the sick, the elderly and the otherwise needy, which induces people to increase their saving for the rainy days, thus depressing effective demand for private consumption.

The growth in consumption is also hampered by slow reform of the Hukou (household registration) system,[8] which constrains the pace of urbanisation and limits the benefits of migrant workers, making them less likely to spend more than is necessary.

---

[4]The term "New Normal Era" used for describing the current economic conditions of China was initially introduced by President Xi Jinping in his speech during the Asia-Pacific Economic Cooperation (APEC) CEO Summit held in Beijing in November 2014 (*Xinhua News Agency* 2014) and further elaborated by Premier Li Keqiang in his Report on the Work of the Government delivered in March 2015 (Li 2015).

[5]Dr Zheng Yongnian, the Director of the East Asian Institute at the National University of Singapore, raised such concern of "New Normal" becoming "Abnormal" in his article published on the Phoenix New Media website (Zheng 2015a).

[6]As reported by *China Daily* USA (*China Daily* 2015).

[7]The data for household final consumption expenditure as a percentage of GDP for China are available from the World Bank, which shows a regressing trend in the household consumption to GDP ratio from the 2000 level of 47% to the 2013 level of 34%.

[8]China's Hukou system, or the household registration system, records officially the identity of any citizen of China as a resident of an area of the country. It has been argued to be a caste system of China, due to its entrenchment of social strata (Perry and Selden 2010).

In terms of export-oriented external demand, the appreciation of the renminbi or yuan against the USD by about 32% in the past decade[9] has put pressure on the growth of foreign demands for goods made in China, while the recent rise in labour cost may further erode the competitiveness of Chinese goods in the international market.

5.2 THE SUPPLY SIDE SOLUTION TO THE DEMAND SIDE CHALLENGES

The central government used to respond to the aforementioned demand side challenges to the Chinese economy by taking a rather short-term approach aimed directly at addressing the demand side problems *per se*, which was exemplified by the humongous stimulus packages launched to counter the powerful global economic headwinds generated by the 2008 GFC. In November 2008, the central government of China launched a stimulus plan consisting of 10 expansionary measures that were accentuated in expediting infrastructure development in agriculture, railways, roads and airports. It was estimated that the implementation of such a plan would cost four trillion yuan or 11.6% of China's GDP in 2009.[10] It was compared to the "New Deal"[11] and considered to be the fastest and largest stimulus plan ever carried out in human economic history in terms of the size of the stimulus packages relative to the size of the economy (Xiao, Wang and Wang 2009).

The plan did work for China. The economic growth momentum was sustained by the stimulus packages, which also produced positive spillover effects outside China, providing the much-needed support for many other countries in Asia and beyond with their economic recovery since 2010 (Arnold 2010). But the government intervention of such an extraordinary scale remains controversial, at least among economists and commentators alike and especially those of the political persuasion akin to the neo-classical fundamentalists. Indeed, some of the concerns voiced by these rather phlegmatic observers are not unreasonable, as the boost to the economy by the stimulus plan was not achieved without incurring certain side effects.

Among which the overly loose credit expansion is a prominent one, as the depressed interest rates helped unleash a blowout of increase in loans that were meant for nurturing industrial capacities and improving their quality and efficiency, but in fact would more readily serve to fuel the asset bubbles in the real estate and stock markets. Even the investment in the real economy made available by such credit expansion could cause problems, as the lion's share of this kind of investment went to fund the projects by the large state-owned enterprises, contributing to the conundrum of excess capacities China now faces.

While the GFC as a Black Swan event carries some element of unpredictability, the ensuing inevitable structural changes in China and the world will be lasting with great ramifications,

[9]The appreciation of the renminbi or yuan against the USD over the last 10 years from 2005 to 2015 is calculated to be around 32.4% by the authors using data from Google Finance.

[10]The GDP of China in 2009 was 34.56 trillion yuan according to the National Bureau of Statistics of China. Thus, a four trillion yuan stimulus plan amounts to 11.6% of China's GDP in 2009.

[11]The four trillion yuan stimulus plan of China was compared to the "New Deal" by Timothy Bond, the Chief Economist of Asia Pacific, Bank of America and Merrill Lynch (*China Quality News Net* 2009).

which necessitates long-term structural solutions. Against such a backdrop, President Xi Jinping has recently announced an overall reform strategy centred around the "supply side economics",[12] of which the fundamental purpose is to improve the overall level of productivity of the economy and to adopt an approach for China's further development whereby the interests of the people must be placed at the centre stage.

As emphasised by President Xi, whilst managing a moderate expansion of aggregate demand, the government will proactively seek to remove excess production capacities, to reduce the level of obsolete inventories, to engage in deleveraging to control the systemic risk inherent in any economy with debts growing in profusion, to cut the overall costs of doing business, and to make up for the weaknesses of the structure of the economy. The thrust of such supply side economic strategy is to strengthen quality supplies, to reduce inefficient supplies, to expand effective supplies, to improve the adaptability and flexibility of the structure of the supply side of the economy, and to improve Total Factor Productivity, thereby making the supply side of the economy well equipped to face up to the challenges brought about by the structural changes in the demand side of the economy.

Such decisive move by the top leadership from the government of China signals the impending tectonic shift in the guiding principles of China's further reform and opening up, from predominantly relying on the demand side short-term stimulus, in which China still has some latitude in making such measures effective given China's still relatively high interest rates level and the availability of the cushioning effect of its sizable foreign exchange reserves, to vigorously carrying out supply side reform that involves long-term, deep and sometimes painful structural adjustments.

However, there is more to the name of such economic strategy than meets the eye, as the saying goes. At the very least, one should not mistake the above-described supply side economics China is going to embrace as something akin to another doctrine of policies that is also associated with the supply side economics, namely the Reaganomics, as they are not at all similar.[13] The latter is in essence about cutting taxes for the corporations and the rich and hope for the trickle-down effects to reach the rest of the society, which has been found to lack an empirical footing as there is little evidence for the benefits of tax cuts for the rich to trickle down (Dabla-Norris *et al.* 2015).

What China will focus on in its own version of the supply side reform will be improving the quality and efficiency of the supplies while streamlining its industries with excess capacities, which involves cutting taxes for the small and micro companies and simplifying regulatory and

---

[12] President Xi Jinping made a detailed exposition on the supply side reform during the Twelfth Meeting of the Central Leading Group on Financial and Economic Affairs on 26th January 2016. More information is available at http://news.xinhuanet.com/fortune/2016-01/27/c_128673404.htm (*Economic Information Daily* 2016).

[13] Unfortunately, some media in the West have already made a wrong comparison between China's supply side economics and the Reaganomics. See, for example, the article "Reagan's Chinese Echo" by *The Economist* available at http://www.economist.com/news/china/21684804-mystery-xi-jinpings-supply-side-strategy-reagans-chinese-echo (*The Economist* 2016).

administrative procedures to reduce the costs and increase the incentives for these companies in their endeavour to be more efficient and innovative.[14]

Moreover, such measures will go hand in hand with boosting effective demand, which entails cutting taxes and enhancing the benefits for the poor, the vulnerable and the disadvantaged members of the society to increase their incomes, thereby making the distribution of income and wealth fairer and more even and achieving the elimination of poverty by 2020 in all rural areas,[15] without being too much concerned about levying a progressive taxation on the rich and the big private or state-owned companies enjoying monopolistic profits.

With such distinctions made and the salient features of the supply side reform of China understood, we should also be aware of the risk of capital flight that is already happening,[16] especially when pursuing a progressive tax scheme is an indispensable component of the people-oriented development strategy. The accompanying trend in the exchange rates of the yuan against the USD and other major currencies carries the risk of over-depreciation of the yuan, which will exacerbate the capital flight and vice versa, hence forming a vicious cycle. While such trend and fluctuations in the exchange rates are worth noting, one should not be overly obsessed about those epiphenomena manifested in the ups and downs of the foreign exchange market. The fundamental solution is again deep and structural, requiring the Chinese economy to upgrade itself to achieve better efficiency, greater productivity and higher competitiveness.

5.3 TOWARDS AN INNOVATION DRIVEN ECONOMY BASED ON PRODUCTIVITY AND HUMAN CAPITAL

Thus, while striving to sustain a medium-high level of growth speed in China under the New Normal Era will be an essential element of policy considerations in the economic management of the country, the most urgent task is to carry out the much-needed structural reform that will help to improve the efficiency and productivity of the economy by upgrading its industries and phasing out excess capacities. Supply side reform that aims to deliver long-term positive structural changes will take a leading role in the overall reform strategy, with short-term stimulus to bolster the aggregate demand only incurred as a last resolve if necessary. A fundamental goal for China to restructure its economy is to build an innovation-driven economy based on productivity

[14]According to Premier Li Keqiang, the reduction in taxes and fees, especially the implementation of tax reduction that is precisely catering for small and micro enterprises, is a positive fiscal policy, which is also conducive for improving the profile of income distribution (Xiao 2015).

[15]President Xi Jinping has recently reiterated the importance of the period of the "Thirteenth Five-Year Plan" in building a comprehensively well-off society, and singled out the elimination of poverty in all rural areas by 2020 as a critical goal that must be achieved by mobilising the resources of the whole nation when necessary (*The Website of Government of China* 2016).

[16]In the one-year period that ended in June 2015, there had been more than US$500 billion worth of capital outflows from China, which coincided with the rather rapid fall in China's foreign exchange reserves from the peak of US$4 trillion to US$3.7 billion during the same period, which was expected to decrease further to US$3.3 trillion by the end of 2015, according to the data and estimation from Citigroup (Sender 2015).

and human capital in its pursuit of long-term sustainable and stable growth at medium-high rates in the next few decades.

Greater policy incentives should be provided to promote innovation-driven growth. As a promising example for China moving into this direction, the authors recently visited Guangdong for field studies and found the government there actively engaged in industrial upgrading and economic restructuring in an attempt to seek new engines of growth that is knowledge based and innovation driven.[17] The importance of those policy initiatives granted, we should be aware that the contribution of the factors underpinning the innovation-driven growth may have its own limit, as in what has been demonstrated empirically as the plausible level of such contribution.

The paradigm economy in innovation-driven growth, the United States, can only grow at around 2% in recent years,[18] despite the fact that the most innovative technology companies and the best universities with the strongest research capabilities in the world are largely concentrated in the country. It is possible that China with a much larger population and greater potential for growth in market size and division of labour can benefit to a greater extent from becoming an innovation-driven economy, but such transition could not happen overnight and we should be prepared to be patient for its positive effects to sink in and intervene if necessary to manage a stable level of increase in the aggregate demand.

The government should emphasise on fostering conditions favourable for the endogenous growth of the economy and support the private sector in their efforts to improve their efficiency and productivity. In particular, as discussed in Chapter 4 of this book, improving agricultural productivity, especially its Total Factor Productivity, is an important avenue towards enhancing the efficiency of the whole economy. With a remarkable human capital profile, China has the potential to do well as a knowledge-based economy if the right conditions are created for its tens of millions of university-educated graduates with the ability to think independently to create, innovate and be entrepreneurial.

The government should also fully capitalise on the benefits and growth potential unleashed by the ongoing urbanisation. China still has plenty of room for further urbanisation.[19] Institutional factors such as social security and the Hukou system need to evolve by having more serious and deeper reforms in order to facilitate a smoother process of urbanisation at the rate that is conducive to sustaining the steady momentum of growth in the New Normal Era. Further urbanisation will unleash new waves of demand for investment in fixed assets and beef up private consumption with newly urbanised population enjoying higher purchasing power as a result of their better-paid occupations found in the cities. The pace of urbanisation and the increase in agricultural

[17]The policy guidelines for Guangdong to upgrade its industries and restructure its economy were summarised by the Party Secretary of Guangdong, Hu Chunhua, as "empty the cage and change the bird" and "be determined to be a reborn phoenix", as reported by *China News Net* (2013).

[18]The data for the GDP growth of the United States are available from the World Bank, according to which the average annual GDP growth rate of the United States is 1.7% from 2000 to 2013.

[19]China's urbanisation rate increased from 17.9% in 1978 to 53.7% in 2013 according to figures from the National Bureau of Statistics of China, but the urbanisation rate calculated based on the criteria of the Hukou system, or the household registration system, was only 35.3% in 2013 (Hu and Chen 2015), indicating the currently incomplete nature of the urbanisation process in China and its huge potential that is yet to be realised.

productivity must go hand in hand to form a virtuous cycle, whereby further urbanisation will boost the demand for agricultural products and help absorb the surplus labour from the agricultural sector with improved productivity, and the abundance of produce from more productive agricultural sector can sufficiently meet the demand of agricultural products in higher quantity and with better quality in the urban areas.

5.4 FUTURE RESEARCH AGENDA

As regarding the impending threat of growth slowdowns, according to the study by the International Monetary Fund (IMF), China seems to be facing no imminent risk of falling into the middle-income trap (Aiyar *et al.* 2013). Given that China is a huge and diverse country with large regional disparities in terms of stages of development and levels of prosperity, analysing China as a whole on the subject of growth slowdowns may lack the granularity required to do justice for such an important topic.

As documented in the literature, the growth prospect of high-technology products as a component of exports by an economy is one of the determinants for whether that economy will encounter growth slowdowns (Eichengreen, Park, and Shin 2013), which is interesting as research on this aspect of growth slowdowns could draw further attention to the alarming trend of slowing growth in exports from China as discussed before and hopefully provide answers to reverse such undesirable trend, thereby helping the economy to move smoothly forward to avoid any imminent and drastic halt in its growth trajectory.

Since the reform and opening up of China starting in the late 1970s, China has experienced stable and rapid GDP growth at a remarkable average annual rate of 9.81% from 1978 to 2014.[20] Meanwhile, China's GDP per capita increased from US$195 to US$3,863 in constant 2005 US$ over the same period.[21] China has also become the biggest trading nation as measured by the sum of exports and imports of goods in 2013 (Anderlin and Hornby 2014). This spectacular growth performance has enabled China to achieve the upper-middle-income status in 2011 (Zhuang, Vandenberg and Huang 2012). However, the growth momentum of China appears to be tapering off in recent years in the aftermath of the 2008 GFC. In the six-year period between 2010 and 2015, China's annual GDP growth has been slowing down from 10.6% in 2010 to under 8% since 2012 and at 6.9% in 2015, the slowest in the last 25 years, according to the National Bureau of Statistics of China.

Furthermore, Mr Lou Jiwei, the Finance Minister of China, sees a greater than 50% chance of the middle-income trap happening in China.[22] In addition, Dr Zheng Yongnian, the Director of the East Asian Institute at the National University of Singapore, contends that China will face difficulties in escaping the middle-income trap (Zheng 2015b). Meanwhile, an empirical

---

[20]The average annual GDP growth rate of China for the period 1978–2014 is calculated based on data from the World Bank, World Development Indicators.

[21]The GDP per capita denoted in 2005 US$ for the period 1978–2014 is obtained from the World Bank, World Development Indicators.

[22]As reported by *China Daily* (2015) USA.

study by Eichengreen, Park and Shin (2012) predicts that China as a whole has a 71% chance of undergoing a slowdown. Up till now, much work and discussion on whether China is or will be stuck in the middle-income trap and hence unable to make the successful transition into the high-income group has focussed on the country as a whole. Given China's regional diversity and variation in terms of Gross Regional Domestic Product (GRDP) per capita, it is more appropriate that the country's possibility of growth slowdown is analysed at the sub-national level. Our future research therefore attempts to fill this gap by studying growth deceleration based on provincial data of mainland China economies as well as that of Hong Kong and Taiwan.

This will be especially relevant for economies relatively strong in exports such as Guangdong and Jiangsu, which are also among the top performers in the Overall Competitiveness in Greater China (Tan *et al.* 2014; Tan, Yuan and Yoong 2015). Whether they will be able to avoid prolonged and significant growth slowdowns have major implications for the future development of Greater China as a whole.

It is therefore of paramount importance for the Chinese government to play the decisive and catalytic role in orchestrating the right kind of competition amongst, as well as development for, its people, economies and regions, in order to ensure stable economic growth at medium-high rates for the next 20–30 years, which will turn China into an innovation-driven economy with its core competitiveness rooted in its abundant stock of quality human capital, prospering with its globally competitive productivity, thereby joining the club of developed countries by the middle of the 21st century.

The issue of whether China can successfully avoid the middle-income trap and continue its development has serious ramifications. China's success will raise the standard of living for close to 1.4 billion people. As a major driver of global growth, a slowdown in China will likely exert great impacts on economic activities in the world, as global economic growth is expected to fall by 0.1% when China's GDP growth decreases by 0.4% (Wei 2015). As China is now at a critical juncture, re-balancing its economy by being less dependent on exports and shifting towards a domestic service-oriented growth strategy, it is timely for us to assess the growth prospects of the Greater China economies *vis-à-vis* other Asian developing economies to better inform policy directions. Our research therefore aims to provide a comprehensive analysis of growth slowdown for 106 Asian developing economies at all income levels. In particular, we seek to investigate whether the determinants of economic slowdown differ across income groups, and estimate the probabilities that the economies in Greater China will experience growth slowdowns in the near future.

In conclusion, it is also pertinent to have a discerning perspective of the current economic slowdown of China in terms of the size of its economy. A simple calculation will show that a 1% GDP growth in the mainland China's economy in 2016 approximately equals to a 2.5% expansion of its economy in 2006. Hence, even if China were to achieve only 6% GDP growth for 2016 and missed its official projected GDP growth target range of 6.5–7%, such an achievement is still equivalent to about 15% increase in the size of its economy a decade ago! Such is the significance of the size of the Chinese economy that would transmit economic waves with implications to the rest of the inter-dependent world.

REFERENCES

Aiyar, Shekhar, Romain Duval, Damien Puy, Yiqun Wu and Longmei Zhang. 2013. *Growth Slowdowns and the Middle-Income Trap*. IMF Working Paper 13/71, International Monetary Fund, Washington DC.

Anderlin, Jamil and Lucy Hornby. 2014. "China Overtakes US as World's Largest Goods Trader." *Financial Times*. January 10. http://www.ft.com/intl/cms/s/0/7c2dbd70-79a6-11e3-b381-00144feabdc0.html#axzz40ySS3UW7.

Arnold, Michael S. 2010. "China's Rebound Eases Slump Fears." *The Wall Street Journal*. September 1. http://online.wsj.com/article/SB10001424052748703882304575464842437541292.html?mod=googlenews_wsj.

Cai, Fang. 2010. "Demographic Transition, Demographic Dividend, and Lewis Turning Point in China." *China Economic Journal* 3(2): 107–119.

China Daily. 2015. "Nation Must Be Alert to Middle-Income Trap." *China Daily USA*. April 15. http://usa.chinadaily.com.cn/epaper/2015-04/28/content_20568239.htm.

China News Net. 2013. "Hu Chunhua Says Guangdong Need to 'Empty the Cage and Change the Bird' and 'Be Determined to Be a Reborn Phoenix.'" *China News Net*. January 17. http://www.chinanews.com/gn/2013/01-17/4497313.shtml.

China Quality News Net. 2009. "The Big Country Mindset in the Financial Crisis." *China Quality News Net*. January 20. http://www.cqn.com.cn/news/zgjyjy/248694.html.

Dabla-Norris, Era, Kalpana Kochhar, Nujin Suphaphiphat, Frantisek Ricka and Evridiki Tsounta. 2015. "Causes and Consequences of Income Inequality : A Global Perspective." *International Monetary Fund*. June 15. http://www.imf.org/external/pubs/cat/longres.aspx?sk=42986.0.

Economic Information Daily. 2016. "Xi Jinping: Strengthen Quality Supplies from the Production Side." *Economic Information Daily*. January 27. http://news.xinhuanet.com/fortune/2016-01/27/c_128673404.htm.

Eichengreen, Barry, Donghyun Park and Kwanho Shin. 2012. "When Fast-Growing Economies Slow Down: International Evidence and Implications for China." *Asian Economic Papers* 11: 42–87.

Eichengreen, Barry, Donghyun Park and Kwanho Shin. 2013. *Growth Slowdowns Redux: New Evidence on the Middle-Income Trap*. NBER Working Paper No. 18673, National Bureau of Economic Research, Cambridge.

Galor, Oded. 2012. "The Demographic Transition: Causes and Consequences." *Cliometrica* 6(1): 1–28.

Hu, Biliang and Chunlai Chen. 2015. "New Urbanisation under Globalisation and the Social Implications in China." *Asia & the Pacific Policy Studies* 2(1): 34–43.

Jenkins, Rhys, Enrique Dussel Peters and Mauricio Mesquita Moreira. 2008. "The Impact of China on Latin America and the Caribbean." *World Development* 36(2): 235–53.

Li, Keqiang. 2015. "Report on the Work of the Government (2015)." *Chinese Central Government*. Third Session of the 12th National People's Congress.

Perry, Elizabeth J. and Mark Selden, eds. 2010. *Chinese Society: Change, Conflict and Resistance*. New York: Routledge.

Sender, Henny. 2015. "Capital Flight Now the Big Concern for Slowing China." *Financial Times*. September 6. http://www.ft.com/intl/cms/s/0/1b7ae16e-4ff7-11e5-8642-453585f2cfcd.html?siteedition=uk#axzz3yoHbPZ81.

Tan, Khee Giap, Randong Yuan and Wei Cher Sangiita Yoong. 2015. *2014 Annual Provincial and Regional Competitiveness Analysis for Greater China Economies: Development Strategies under the New Normal*. Singapore: World Scientific Publishing Company.

Tan, Khee Giap, Randong Yuan, Wei Cher Sangiita Yoong and Mu Yang. 2014. *Annual Analysis of Competitiveness, Simulation Studies and Development Perspective for 34 Greater China Economies: 2000–2010*. Singapore: World Scientific Publishing Company.

Tan, Kong Yam, Tilak Abeysinghe and Khee Giap Tan. 2015. "Shifting Drivers of Growth: Policy Implications for ASEAN-5." *Asian Economic Papers* 14(1): 157–173.

Tao, Ran and Zhigang Xu. 2007. "Urbanization, Rural Land System and Social Security for Migrants in China." *Journal of Development Studies* 43(7): 1301–1320.

The Economist. 2016. "Reagan's Chinese Echo: The Mystery of Xi Jinping's Supply-Side Strategy." *The Economist*. January 2. http://www.economist.com/news/china/21684804-mystery-xi-jinpings-supply-side-strategy-reagans-chinese-echo.

The Website of Government of China. 2016. "Xi Jinping: Make Sure That All Rural Poor Population Are out of Poverty by 2020." *The Website of Government of China*. January 30. http://news.china.com/domesticgd/10000159/20160130/21367402_1.html.

Wei, Shang Jin. 2015. "PRC's Slowdown for Asia's Growth – Key Takeaways." *Asian Development Blog*. July 25. http://blogs.adb.org/blog/prc-s-slowdown-asia-s-growth-key-takeaways.

Xiao, Nan. 2015. "Li Keqiang: Reduction in Taxes and Fees Is by Itself a Positive Fiscal Policy." *People's Daily Online*. February 26. http://politics.people.com.cn/n/2015/0226/c1001-26598022.html.

Xiao, Yingying, Yun Wang and Jing Wang. 2009. "Sustainable Recovery Is the Key." *Economic Information Daily*. May 11 http://jjckb.xinhuanet.com/gjxw/2009-11/05/content_189364.htm.

Xinhua News Agency. 2014. "Chinese President Proposes Asia-Pacific Dream." *Xinhua News Agency*. September 11 http://www.apec-china.org.cn/41/2014/11/09/3@2418.htm.

Zhang, Xiaobo, Jin Yang and Shenglin Wang. 2011. "China Has Reached the Lewis Turning Point." *China Economic Review* 22(4): 542–554.

Zheng, Yongnian. 2015a. "Prevent the 'New Normal' from Becoming the 'Abnormal.'" *Phoenix New Media*. April 29 http://news.ifeng.com/exclusive/lecture/special/zhengyongnian02/.

Zheng, Yongnian. 2015b. "中国难逃中等收入陷阱 China Will Face Difficulty in Espacing the Middle Income Trap." *Caixin.com*. April 2. http://finance.ifeng.com/a/20150402/13603135_0.shtml.

Zhuang, Juzhong, Paul Vandenberg and Yiping Huang. 2012. *Growing beyond the Low-Cost Advantage: How the People's Republic of China Can Avoid the Middle-Income Trap*. Mandaluyong City: Asian Development Bank. http://www.adb.org/sites/default/files/publication/30036/growing-beyond-prc-avoid-middle-income-trap.pdf.

Appendix 1

Provincial Level Economies Competitiveness Analysis — List of Indicators

No.	Indicator	Unit	Reversed Indicator
1	**Macroeconomic Stability**		
1.1	***Regional Economic Vibrancy***		
1.1.01	Gross Regional Product	100 Million Yuan, Constant Prices at 2000	
1.1.02	Real GRP Growth	Percentage Change per Annum	
1.1.03	GRP Per Capita	Yuan per person, Constant Prices at 2000	
1.1.04	Primary Industry	100 Million Yuan, Constant Prices at 2000	
1.1.05	Secondary Industry	100 Million Yuan, Constant Prices at 2000	
1.1.06	Tertiary Industry	100 Million Yuan, Constant Prices at 2000	
1.1.07	Gross Capital Formation (%)	Percentage of GRP	
1.1.08	Inflation Rate (Consumer Price Index)	Percentage Change per Annum	*R*
1.1.09	Inflation Rate (Retail Price Index)	Percentage Change per Annum	*R*
1.1.10	Net Regional Trade	100 Million Yuan, Constant Prices at 2000	
1.1.11	Regional Tourist Arrivals	10,000 Person-Times	
1.1.12	Regional Tourism Receipts	100 Million Yuan, Constant Prices at 2000	
1.1.13	Planned Regional Investment	100 Million Yuan, Constant Prices at 2000	
1.1.14	Realized Regional Investment	100 Million Yuan, Constant Prices at 2000	
1.1.15	Industry Gross Output Value	100 Million Yuan, Constant Prices at 2000	
1.2	***Openness To Trade and Services***		
1.2.01	Trade Balance (%)	Percentage of GRP	
1.2.02	International Merchandise Exports	100 Million Yuan, Constant Prices at 2000	
1.2.03	International Merchandise Imports	100 Million Yuan, Constant Prices at 2000	

(*Continued*)

(*Continued*)

No.	Indicator	Unit	Reversed Indicator
1.2.04	Openness To Trade	Ratio	
1.2.05	International Visitor Arrivals	10,000 Person-Times	
1.2.06	International Tourism Receipts	100 Million Yuan, Constant Prices at 2000	
1.3	***Attractiveness To Foreign Investors***		
1.3.01	Planned Foreign Direct Investment (Excl. Regional Investment)	100 Million Yuan, Constant Prices at 2000	
1.3.02	Realized Foreign Direct Investment (Excl. Regional Investment)	100 Million Yuan, Constant Prices at 2000	
1.3.03	Foreign Funded Enterprises (%) (Industrial Enterprises Above Designated Size)	Percentage of Total Enterprises	
1.3.04	Total Taxes on Foreign Funded Companies	10,000 Yuan, Constant Prices at 2000	
2	**Government and Institutional Setting**		
2.1	***Government Policies and Fiscal Sustainability***		
2.1.01	Government Revenue	100 Million Yuan, Constant Prices at 2000	
2.1.02	Tax Revenue	100 Million Yuan, Constant Prices at 2000	
2.1.03	Tax Revenue/Government Revenue	Percentage of Government Revenue	
2.1.04	Government Consumption Expenditure	Percentage of GRP	
2.1.05	Budget Balance	Percentage of GRP	
2.2	***Institutions, Governance and Leadership***		
2.2.01	Domestic Funded Enterprises (Industrial Enterprises Above Designated Size)	Number	
2.2.02	State-Owned Enterprises (Industrial Enterprises Above Designated Size)	Percentage of Total Enterprises	*R*
2.2.03	Collective-Owned Enterprises (Industrial Enterprises Above Designated Size)	Percentage of Total Enterprises	*R*
2.2.04	Limited Liability Corporations (Industrial Enterprises Above Designated Size)	Percentage of Total Enterprises	
2.2.05	Joint-Stock Companies (Industrial Enterprises Above Designated Size)	Percentage of Total Enterprises	
2.2.06	Private Enterprises (Industrial Enterprises Above Designated Size)	Percentage of Total Enterprises	
2.2.07	Foreign Funded Enterprises (Industrial Enterprises Above Designated Size)	Number	
3	**Financial, Businesses and Manpower Conditions**		
3.1	***Financial Deepening and Business Efficiency***		
3.1.01	Household Savings Deposits	100 Million Yuan, Constant Prices at 2000	
3.1.02	Total Deposits	100 Million Yuan, Constant Prices at 2000	
3.1.03	Total Loans	100 Million Yuan, Constant Prices at 2000	

(*Continued*)

(*Continued*)

No.	Indicator	Unit	Reversed Indicator
3.2	***Labour Market Flexibility***		
3.2.01	Total Employed Persons	10,000 persons	
3.2.02	Self-Employed Persons (Urban Area)	Percentage of Total Employment	
3.2.03	Unemployment Rate (Urban Area)	Percentage	*R*
3.2.04	Secondary & Tertiary Staff Workers (%)	Percentage of Total Employment	
3.2.05	Average Annual Wage of Staff and Workers	Yuan, Constant Prices at 2000	
3.2.06	Total Wages Bill of Employed Persons in Urban Units	100 Million Yuan, Constant Prices at 2000	
3.2.07	Wages Paid Out by State-Owned Units	Percentage of Total Wages	*R*
3.2.08	Wages Paid Out by Urban Collective Units	Percentage of Total Wages	*R*
3.2.09	Wages Paid Out by Other Units	Percentage of Total Wages	
3.3	***Productivity Performance***		
3.3.01	Overall Labour Productivity	Yuan per Person-Year, Constant Prices at 2000	
3.3.02	Primary Industry, Value-Added per Worker	Yuan per Person-Year, Constant Prices at 2000	
3.3.03	Secondary Industry, Value-Added per Worker	Yuan per Person-Year, Constant Prices at 2000	
3.3.04	Tertiary Industry, Value-Added per Worker	Yuan per Person-Year, Constant Prices at 2000	
3.3.05	Investment in Fixed Assets	100 Million Yuan, Constant Prices at 2000	
4	**Quality of Life and Infrastructure Development**		
4.1	***Physical Infrastructure***		
4.1.01	Population	10,000 Persons, Year-end	
4.1.02	Urban Population	Percentage of Total Population	
4.1.03	Population Density	Persons per Square Kilometre	*R*
4.1.04	Length of Railways in Operation	Kilometre	
4.1.05	Length of Highways	Kilometre	
4.1.06	Length of Navigable Inland Waterways	Kilometre	
4.1.07	Passenger Distance of Railways in Operation	100 million person-kilometre	
4.1.08	Passenger Distance of Highways	100 million person-kilometre	
4.1.09	Passenger Distance of Navigable Inland Waterways	100 million person-kilometre	
4.1.10	Passenger Distance of Total Civil Aviation Routes	100 million person-kilometre	
4.1.11	Motor Vehicles per km of Highways	Per Kilometre	*R*

(*Continued*)

(Continued)

No.	Indicator	Unit	Reversed Indicator
4.1.12	Coal	10,000 tonnes	
4.1.13	Oil	Tonne	
4.1.14	Electric Hydropower	100 million kilowatt hour	
4.1.15	Crude Steel	10,000 tonnes	
4.1.16	Pig Iron	10,000 tonnes	
4.1.17	Steel Products	10,000 tonnes	
4.1.18	Cement	10,000 tonnes	
4.1.19	Total Land Area	Square Kilometre	
4.1.20	Cultivated Land	1,000 Ha	
4.2	***Technological Infrastructure***		
4.2.01	Persons per Fixed Telephone Subscriber	Per Fixed Telephone Subscriber	*R*
4.2.02	Internet Penetration	Percentage of Total Population	
4.2.03	Personal Computer	Per 100 Urban Households	
4.2.04	High Technology Expenditure	Percentage of Total Government Expenditure	
4.2.05	Persons per Mobile Telephone Subscriber	Per Mobile Telephone Subscriber	*R*
4.2.06	Urban Television Sets	Units per 100 Households	
4.2.07	Rural Television Sets	Units per 100 Households	
4.2.08	Expenditure on Science and Technology	Percentage of GRP	
4.2.09	Number of Scientific and Technical Journal Publications	Number	
4.2.10	Patents Granted	Number	
4.3	***Standard of Living, Education and Social Stability***		
4.3.01	Illiteracy Rate	Percentage of Population Age 15 and Above	*R*
4.3.02	Number of Educational Institutions	Total Units	
4.3.03	Disparity between Primary and Secondary Institutions	Percentage	*R*
4.3.04	Tertiary Institutions	Percentage of Total Institutions	
4.3.05	Student-Teacher Ratio (Primary)	Ratio	*R*
4.3.06	Student-Teacher Ratio (Secondary)	Ratio	*R*
4.3.07	Student-Teacher Ratio (Tertiary)	Ratio	*R*
4.3.08	Health Expenditure	Percentage of Total Government Expenditure	
4.3.09	Urban Population Growth	Percentage Change per Annum	

(*Continued*)

(Continued)

No.	Indicator	Unit	Reversed Indicator
4.3.10	Air Pollution	milligram/cu.m	*R*
4.3.11	Ageing Profile	Percentage of Population Over 65 years	*R*
4.3.12	Old Age Dependency Ratio	Ratio	*R*
4.3.13	Motor Vehicles per Person	Per Person	
4.3.14	Death Rate	Per 1,000 Persons	*R*
4.3.15	Adequacy of Hospitals	10,000 Persons per Hospital	*R*
4.3.16	Population per Hospital Bed	10,000 Persons per Hospital Bed	*R*
4.3.17	Population per Medical Personnel	10,000 Persons per Medical Personnel	*R*
4.3.18	Per Capita Annual Urban Household Income	Yuan, Constant Prices at 2000	

Appendix 2

Competitiveness Analysis — The Algorithm

A step-by-step description of the ranking process is described below for N economies, M practical indicators and C environments, with each environment comprising S sub-environments.

(1) Compute the mean value of practical indicator j ($j = 1, \ldots, M$),

$$\bar{X}_j = \frac{1}{N}\sum_{i=1}^{N} X_{ij},$$

where X_{ij} represents the value that economy i ($i = 1, \ldots, N$) takes for practical indicator j.

(2) For each practical indicator j ($j = 1, \ldots, M$), calculate its standard deviation (SD),

$$\text{SD}_j = \sqrt{\frac{1}{N}\sum_{i=1}^{N}(X_{ij} - \bar{X}_j)^2}.$$

(3) Compute the standardized value of indicator (SVI) that each economy i ($i = 1, \ldots, N$) takes under each of the practical indicators j ($j =1, \ldots, M$),

$$\text{SVI}_{ij} = \frac{X_{ij} - \bar{X}_j}{SD_j}.$$

(4) Compute the "ranked" standardized value of indicator (RSVI) that each economy i ($i = 1, \ldots, N$) takes under each of the practical indicators j ($j = 1, \ldots, M$):

$$\text{RSVI}_{ij} = \begin{cases} SVI_{ij}, \text{if a higher value better} \\ -SVI_{ij}, \text{if a lower value better} \end{cases}.$$

(5) For each of the practical indicators j ($j = 1, \ldots, M$), a ranking can be obtained for economies: economies with a higher value of RSVI for indicator j are ranked ahead of those with a lower value.

(6) For each economy i ($i = 1, \ldots, N$), calculate the RSVI for each sub-environment k ($k = 1, \ldots, S$) belonging to environment l ($l = 1, \ldots, C$),

$$\text{Raw\_RSVI}_{i,lk} = \frac{1}{y_{lk}} \sum_{p=1}^{y_{lk}} \text{RSVI}_{i,Jlk,p},$$

$$\text{Mean\_RSVI}_{lk} = \frac{1}{N} \sum_{i=1}^{N} \text{Raw\_RSVI}_{i,lk},$$

$$\text{SD\_RSVI}_{lk} = \sqrt{\frac{1}{N} \sum_{i=1}^{N} \left(\text{Raw\_RSVI}_{i,lk} - \text{Mean\_RSVI}_{lk}\right)^2},$$

$$\text{RSVI}_{i,lk} = \frac{\text{Raw\_RSVI}_{i,lk} - \text{Mean\_RSVI}_{lk}}{\text{SD\_RSVI}_{lk}},$$

where y_{lk} is the total number of practical indicators under sub-environment k of environment l and (RSVIi, *jlk*, 1, . . . , RSVIi, *jlk*, *ylk*) are the RSVIs for economy i that make up sub-environment k of environment l.

(7) For each economy i ($i = 1, \ldots, N$), calculate the RSVI for each environment l ($l = 1, \ldots, C$),

$$\text{Raw\_RSVI}_{i,l} = \frac{1}{S_l} \sum_{k=1}^{S_l} \text{RSVI}_{i,lk},$$

$$\text{Mean\_RSVI}_{l} = \frac{1}{N} \sum_{i=1}^{N} \text{Raw\_RSVI}_{i,l},$$

$$\text{SD\_RSVI}_{l} = \sqrt{\frac{1}{N} \sum_{i=1}^{N} \left(\text{Raw\_RSVI}_{i,l} - \text{Mean\_RSVI}_{l}\right)^2},$$

$$\text{RSVI}_{i,l} = \frac{\text{Raw\_RSVI}_{i,l} - \text{Mean\_RSVI}_{l}}{\text{SD\_RSVI}_{l}},$$

where (RSVIi, *l*1, . . . , RSVIi, *lS*) are the RSVIs for the S sub-environments under each environment l.

(8) Overall rank score of economy i ($i = 1, \ldots, N$),

$$\text{Raw\_}R_i = \frac{1}{C} \sum_{l=1}^{C} \text{RSVI}_{i,l},$$

$$\text{Mean\_}R = \frac{1}{N} \sum_{i=1}^{N} \text{Raw\_}R_i,$$

$$\text{SD\_}R = \sqrt{\frac{1}{N} \sum_{i=1}^{N} \left(\text{Raw\_}R_i - \text{Mean\_}R\right)^2},$$

$$R_i = \frac{\text{Raw\_}R_i - \text{Mean\_}R}{\text{SD\_}R}.$$

Economies with a higher *Ri* are ranked ahead of those with lower value of *Ri*, and the economy with the highest *Ri* is the most competitive economy.

Step (5) provides the ranking of each economy for each individual practical indicator. To achieve this ranking, Step (4) adjusts the value of the SVIs so that a higher value will lead to a better ranking in terms of "competitiveness". Depending on the nature of the indicator in question, a higher or lower value may reflect a more "competitive" economy. Take for instance the practical indicators "Gross Regional Product (GRP)" and "Unemployment Rate (Urban Area)". A higher GRP but a lower unemployment rate suggests better economic performance which makes an economy more competitive. In most cases where a higher value is better (e.g. GRP), the SVIs of economies are considered, and those with a higher SVI value will have a better ranking. However, for indicators where the inverse is true (e.g. Unemployment Rate (Urban Area)), the "negative SVI" values are compared between economies and a lower SVI value will lead to a better ranking. Step (4) thus seeks to make all standardizes values of all practical indicators consistent for ranking purposes.

Step (6) determines the sub-environment rankings of each economy. The average RSVI of all the indicators in the sub-environment are calculated, re-standardised across all the economies and compared to other economies. Economies with a higher average RSVI rank better in the sub-environment.

To arrive at the ranking for each environment, the RSVIs of the sub-environments are averaged and re-standardised across all the economies as detailed by Step (7). Finally, Step (8) requires the RSVI values of each environment to be averaged and re-standardised across all the economies to determine the overall ranking of the economy. Economies with a higher RSVI are ranked ahead of those with a lower RSVI.

Although the number of sub-environments and indicators varies for each main environment, the aggregate score for each main environment is given an equivalent weighting: 25% of the Greater China Overall Competitiveness Index. Identical weights are assigned to each environment as they represent equivalent significance to the computation of the Index. This method is repeated and applied consistently across all the economies to ensure precision of the rankings. Mathematically, this can be illustrated as follows:

Greater China Overall Competitiveness Index:

= 25% × (Macroeconomic Stability)
+ 25% × (Government and Institutional Setting)
+ 25% × (Financial, Businesses and Manpower Conditions)
+ 25% × (Quality of Life and Infrastructure Development).

Appendix 3

Regional Competitiveness Analysis — List of Indicators and Notes on Data Aggregation Method

No.	Indicator^	Unit	Reversed Indicator	Aggregation Method*
1	**Macroeconomic Stability**			
1.1	***Regional Economic Vibrancy***			
1.1.01	Gross Regional Product	100 Million Yuan, Constant Prices at 2000		Method 1
1.1.02	Real GRP Growth	Percentage Change per Annum		Method 2
1.1.03	GRP Per Capita	Yuan per person, Constant Prices at 2000		Method 3
1.1.04	Primary Industry	100 Million Yuan, Constant Prices at 2000		Method 1
1.1.05	Secondary Industry	100 Million Yuan, Constant Prices at 2000		Method 1
1.1.06	Tertiary Industry	100 Million Yuan, Constant Prices at 2000		Method 1
1.1.07	Gross Capital Formation (%)	Percentage of GRP		Method 3
1.1.08	Inflation Rate (Consumer Price Index)	Percentage Change per Annum	*R*	Method 4a
1.1.09	Inflation Rate (Retail Price Index)	Percentage Change per Annum	*R*	Method 4a
1.1.10	Industry Gross Output Value	100 Million Yuan, Constant Prices at 2000		Method 1
1.2	***Openness To Trade and Services***			
1.2.01	Trade Balance (%)	Percentage of GRP		Method 3
1.2.02	International Merchandise Exports	100 Million Yuan, Constant Prices at 2000		Method 1
1.2.03	International Merchandise Imports	100 Million Yuan, Constant Prices at 2000		Method 1

(Continued)

(Continued)

No.	Indicator^	Unit	Reversed Indicator	Aggregation Method*
1.2.04	Openness To Trade	Ratio		Method 3
1.2.05	International Visitor Arrivals	10,000 Person-Times		Method 1
1.2.06	International Tourism Receipts	100 Million Yuan, Constant Prices at 2000		Method 1
1.3	***Attractiveness To Foreign Investors***			
1.3.01	Planned Foreign Direct Investment (Excl. Regional Investment)	100 Million Yuan, Constant Prices at 2000		Method 1
1.3.02	Realized Foreign Direct Investment (Excl. Regional Investment)	100 Million Yuan, Constant Prices at 2000		Method 1
1.3.03	Foreign Funded Enterprises (%) (Industrial Enterprises Above Designated Size)	Percentage of Total Enterprises		Method 3
1.3.04	Total Taxes on Foreign Funded Companies	10,000 Yuan, Constant Prices at 2000		Method 1
2	**Government and Institutional Setting**			
2.1	***Government Policies and Fiscal Sustainability***			
2.1.01	Government Revenue	100 Million Yuan, Constant Prices at 2000		Method 1
2.1.02	Tax Revenue	100 Million Yuan, Constant Prices at 2000		Method 1
2.1.03	Tax Revenue/Government Revenue	Percentage of Government Revenue		Method 3
2.1.04	Government Consumption Expenditure	Percentage of GRP		Method 3
2.1.05	Budget Balance	Percentage of GRP		Method 3
2.2	***Institutions, Governance and Leadership***			
2.2.01	Domestic Funded Enterprises (Industrial Enterprises Above Designated Size)	Number		Method 1
2.2.02	State-Owned Enterprises (Industrial Enterprises Above Designated Size)	Percentage of Total Enterprises	*R*	Method 3
2.2.03	Collective-Owned Enterprises (Industrial Enterprises Above Designated Size)	Percentage of Total Enterprises	*R*	Method 3

(Continued)

(Continued)

No.	Indicator^	Unit	Reversed Indicator	Aggregation Method*
2.2.04	Limited Liability Corporations (Industrial Enterprises Above Designated Size)	Percentage of Total Enterprises		Method 3
2.2.05	Joint-Stock Companies (Industrial Enterprises Above Designated Size)	Percentage of Total Enterprises		Method 3
2.2.06	Private Enterprises (Industrial Enterprises Above Designated Size)	Percentage of Total Enterprises		Method 3
2.2.07	Foreign Funded Enterprises (Industrial Enterprises Above Designated Size)	Number		Method 1
3	**Financial, Businesses and Manpower Conditions**			
3.1	***Financial Deepening and Business Efficiency***			
3.1.01	Household Savings Deposits	100 Million Yuan, Constant Prices at 2000		Method 1
3.1.02	Total Deposits	100 Million Yuan, Constant Prices at 2000		Method 1
3.1.03	Total Loans	100 Million Yuan, Constant Prices at 2000		Method 1
3.2	***Labour Market Flexibility***			
3.2.01	Total Employed Persons	10,000 persons		Method 1
3.2.02	Self-Employed Persons (Urban Area)	Percentage of Total Employment		Method 3
3.2.03	Unemployment Rate (Urban Area)	Percentage	*R*	Method 3
3.2.04	Secondary & Tertiary Staff Workers (%)	Percentage of Total Employment		Method 3
3.2.05	Average Annual Wage of Staff and Workers	Yuan, Constant Prices at 2000		Method 4a
3.2.06	Total Wages Bill of Employed Persons in Urban Units	100 Million Yuan, Constant Prices at 2000		Method 1
3.2.07	Wages Paid Out by State-Owned Units	Percentage of Total Wages	*R*	Method 3
3.2.08	Wages Paid Out by Urban Collective Units	Percentage of Total Wages	*R*	Method 3
3.2.09	Wages Paid Out by Other Units	Percentage of Total Wages		Method 3

(Continued)

(Continued)

No.	Indicator^	Unit	Reversed Indicator	Aggregation Method*
3.3	***Productivity Performance***			
3.3.01	Overall Labour Productivity	Yuan per Person-Year, Constant Prices at 2000		Method 3
3.3.02	Primary Industry, Value-Added per Worker	Yuan per Person-Year, Constant Prices at 2000		Method 3
3.3.03	Secondary Industry, Value-Added per Worker	Yuan per Person-Year, Constant Prices at 2000		Method 3
3.3.04	Tertiary Industry, Value-Added per Worker	Yuan per Person-Year, Constant Prices at 2000		Method 3
3.3.05	Investment in Fixed Assets	100 Million Yuan, Constant Prices at 2000		Method 1
4	**Quality of Life and Infrastructure Development**			
4.1	***Physical Infrastructure***			
4.1.01	Population	10,000 Persons, Year-end		Method 1
4.1.02	Urban Population	Percentage of Total Population		Method 3
4.1.03	Population Density	Persons per Square Kilometre	*R*	Method 3
4.1.04	Length of Railways in Operation	Kilometre		Method 1
4.1.05	Length of Highways	Kilometre		Method 1
4.1.06	Length of Navigable Inland Waterways	Kilometre		Method 1
4.1.07	Passenger Distance of Railways in Operation	100 million person-kilometre		Method 1
4.1.08	Passenger Distance of Highways	100 million person-kilometre		Method 1
4.1.09	Passenger Distance of Navigable Inland Waterways	100 million person-kilometre		Method 1
4.1.10	Passenger Distance of Total Civil Aviation Routes	100 million person-kilometre		Method 1
4.1.11	Motor Vehicles per km of Highways	Per kilometre	*R*	Method 3
4.1.12	Coal	10,000 tonnes		Method 1
4.1.13	Oil	Tonne		Method 1
4.1.14	Electric Hydropower	100 million kilowatt hour		Method 1
4.1.15	Crude Steel	10,000 tonnes		Method 1
4.1.16	Pig Iron	10,000 tonnes		Method 1
4.1.17	Steel Products	10,000 tonnes		Method 1
4.1.18	Cement	10,000 tonnes		Method 1

(Continued)

(*Continued*)

No.	Indicator^	Unit	Reversed Indicator	Aggregation Method*
4.1.19	Total Land Area	Square Kilometre		Method 1
4.1.20	Cultivated Land	1,000 Ha		Method 1
4.2	***Technological Infrastructure***			
4.2.01	Persons per Fixed Telephone Subscriber	Per Fixed Telephone Subscriber	*R*	Method 3
4.2.02	Internet Penetration	Percentage of Total Population		Method 3
4.2.03	Personal Computer	Per 100 Urban Households		Method 4a
4.2.04	High Technology Expenditure	Percentage of Total Government Expenditure		Method 3
4.2.05	Persons per Mobile Telephone Subscriber	Per Mobile Telephone Subscriber	*R*	Method 3
4.2.06	Urban Television Sets	Units per 100 Households		Method 4a
4.2.07	Rural Television Sets	Units per 100 Households		Method 4a
4.2.08	Expenditure on Science and Technology	Percentage of GRP		Method 3
4.2.09	Number of Scientific and Technical Journal Publications	Number		Method 1
4.2.10	Patents Granted	Number		Method 1
4.3	***Standard of Living, Education and Social Stability***			
4.3.01	Illiteracy Rate	Percentage of Population Age 15 and Above	*R*	Method 3
4.3.02	Number of Educational Institutions	Total Units		Method 1
4.3.03	Disparity between Primary and Secondary Institutions	Percentage	*R*	Method 4a
4.3.04	Tertiary Institutions	Percentage of Total Institutions		Method 3
4.3.05	Student–Teacher Ratio (Primary)	Ratio	*R*	Method 4a
4.3.06	Student–Teacher Ratio (Secondary)	Ratio	*R*	Method 4a
4.3.07	Student–Teacher Ratio (Tertiary)	Ratio	*R*	Method 4a
4.3.08	Health Expenditure	Percentage of Total Government Expenditure		Method 3
4.3.09	Urban Population Growth	Percentage Change per Annum		Method 3
4.3.10	Air Pollution	milligram/cu.m	*R*	Method 4b
4.3.11	Ageing Profile	Percentage of Population Over 65 years	*R*	Method 3

(*Continued*)

(Continued)

No.	Indicator^	Unit	Reversed Indicator	Aggregation Method*
4.3.12	Old Age Dependency Ratio	Ratio	*R*	Method 3
4.3.13	Motor Vehicles per Person	Per Person		Method 3
4.3.14	Death Rate	Per 1,000 Persons	*R*	Method 3
4.3.15	Adequacy of Hospitals	10,000 Persons per Hospital	*R*	Method 3
4.3.16	Population per Hospital Bed	10,000 Persons per Hospital Bed	*R*	Method 3
4.3.17	Population per Medical Personnel	10,000 Persons per Medical Personnel	*R*	Method 3
4.3.18	Per Capita Annual Urban Household Income	Yuan, Constant Prices at 2000		Method 4a

Note: ^ The five indicators, namely 1.1.10 Net Regional Trade, 1.1.11 Regional Tourist Arrivals, 1.1.12 Regional Tourism Receipts, 1.1.13 Planned Regional Investment, and 1.1.14 Realized Regional Investment, are used for the competitiveness ranking analysis by economy, but they are not used for the competitiveness ranking analysis by region. This is due to the nature of their definition, which renders it impossible to use sub-regional data alone to derive the regional data, and also due to unavailability of relevant data published on the regions.

***Method 1: Since it is based on absolute quantity, we can get the regional value by summing across all the economies within this region.**

Example for Method 1: Gross Regional Domestic Product (GRDP); Unit: 100 million yuan, Constant Prices at 2000

For Region *A* with *n* economies at time *t*,

$$\text{GRDP}_{A,t} = \sum_{i=1}^{n} GRDP_{i,t}.$$

Method 2: Once we have computed the aggregated data for each region across the years according to Method 1, it is straightforward to calculate say for example growth rate of that particular data by region according to its definition.

Example for Method 2: GRDP Growth; Unit: Percentage Change Per Annum

For Region *A* with *n* economies at time *t*,

$$\text{GRDP Growth}_{A,t} = \frac{\text{GRDP}_{A,t} - \text{GRDP}_{A,t-1}}{\text{GRDP}_{A,t-1}}.$$

Method 3: Since both the numerator and denominator are absolute quantities, we aggregate them across all the economies within the region to get the regional values, from which we compute the ratio, hence arriving at the regional value for indicator.

Example for Method 3: Gross Capital Formation; Unit: Percentage of GRDP

For Region *A* with *n* economies at time *t*,

$$\text{Gross Capital Formation}_{A,t} = \frac{\sum_{i=1}^{n} \text{Gross Capital Formation}_{i,t}}{\sum_{i=1}^{n} \text{GRDP}_{i,t}}.$$

Method 4: Since the economy-level data is in the form of ratio or percentage that cannot be easily disaggregated into numerator and denominator, the regional value for the indicator is obtained by taking the weighted average of economy-level data. We can either take the weighted average based on population (Method 4a) or weighted average based on land size (Method 4b), depending on the nature of the indicator.

Example for Method 4: Inflation Rate (Consumer Price Index); Unit: Percentage Change per Annum

For Region *A* with n economies at time *t*,

$$\text{Inflation Rate}_{A,t} = \frac{\sum_{i=1}^{n} \left(\text{Population}_{i,t} \times \text{Inflation Rate}_{i,t}\right)}{\sum_{i=1}^{n} \text{Population}_{i,t}}.$$